PLANT TISSUE CULTURE FOR HORTICULTURAL CROPS

THE AUTHORS

Dr. Javid Iqbal Mir is Principal Scientist (Agricultural Biotechnology) at ICAR-Central Institute of Temperate Horticulture (CITH), Srinagar, and Heads both the Clean Plant Centre and the Vegetable Science Division. He joined ICAR in 2005 as an ARS Scientist at ICAR-CRIJAF, Kolkata. Over the past 20 years, he has led impactful research and administrative initiatives in temperate horticulture.

He established advanced biotech and quarantine facilities at Srinagar, Mukteshwar, and Dirang, and has handled 45+ research projects funded by ICAR, DBT, DST, MIDH, NHB, PPV&FRA, JICA, and others. He has strong collaborations with institutes like IARI, NBPGR, IIHR, SKUAST-K, University of Kashmir, and IIT Roorkee.

Dr. Mir has developed and released 17 varieties of apple, walnut, cherry, apricot, onion, and garlic, secured 3 design patents, and obtained 5 PPV&FRA protections. He has published 200+ research papers, 15 books, 40 extension bulletins, and 100+ book chapters. He is also actively involved in technology dissemination through TSP, Mera Gaon Mera Gaurav, NICRA, and NMSHE programs.

Administratively, he has held key roles including In-Charge PME, ITMU, AO, DDO, and Chair of DPAC, and has overseen platforms like IPR, PIMS, PERMISNET, and HYPM. His extension efforts include four success stories where farmers doubled income by adopting his technologies.

A recipient of multiple awards including Best Researcher and Outstanding Achievement Award, Dr. Mir is editor for several journals and has mentored 40+ MSc and PhD scholars. He remains a leading voice in plant biotechnology and temperate crop improvement in India.

Dr. Sadiah Shafi holds a Ph.D. in Genetics and Plant Breeding from Sher-e-Kashmir University of Agricultural Sciences and Technology of Kashmir (SKUAST-K). She is currently working as a Research Professional at the ICAR-Central Institute of Temperate Horticulture (ICAR-CITH), Srinagar, and has served as a Visiting Research Scholar at Kansas State University, USA, where she contributed to multi-omics research on climate-induced stress in crops.

Dr. Shafi's research expertise encompasses plant genetics, molecular breeding, functional genomics, and high-throughput phenotyping, with hands-on experience in leading laboratories across India and the United States. She has authored 40+ research publications and book chapters.

A consistently top-ranked scholar and university gold medalist, Dr. Shafi has received several awards, fellowships, and best paper/poster recognitions for her scientific contributions. Passionate about innovation in plant science, Dr. Shafi is committed to advancing global sustainable crop improvement and mentoring the next generation of plant researchers.

PLANT TISSUE CULTURE FOR HORTICULTURAL CROPS

– Authored by –

Dr. Javid Iqbal Mir

Dr. Sadiah Shafi

– Complied by –

Mohd. Ashaq Kuchay	Nusrat Perveen	Salwee Yasmin
Farhana Hassan	Uzma Rashid	Syed Mudassir Jeelani
Ashiq Hussain Lone	Mohammad Irfan	Jyoti Priya
Reena Prusty	Kavita. R.	Mohd. Abbas Shah
Sajad Un Nabi	Om Chand Sharma	Waseem Hassan Raja
Geetika Malik	Mahendra Kumar Verma	

2026

Daya Publishing House®

A Division of

Astral International Pvt. Ltd.

New Delhi – 110 002

ISBN: 978-93-5919-872-9

Publisher's Note:

Every possible effort has been made to ensure that the information contained in this book is accurate at the time of going to press, and the publisher and author cannot accept responsibility for any errors or omissions, however caused. No responsibility for loss or damage occasioned to any person acting, or refraining from action, as a result of the material in this publication can be accepted by the editor, the publisher or the author. The Publisher is not associated with any product or vendor mentioned in the book. The contents of this work are intended to further general scientific research, understanding and discussion only. Readers should consult with a specialist where appropriate.

Every effort has been made to trace the owners of copyright material used in this book, if any. The author and the publisher will be grateful for any omission brought to their notice for acknowledgement in the future editions of the book.

Published by : **Daya Publishing House®**
A Division of
Astral International Pvt. Ltd.
– ISO 9001:2015 Certified Company –
4736/23, Ground Floor, Ansari Road,
Darya Ganj, New Delhi-110 002
Ph. 011-43549197, 8130496929
E-mail: info@astralint.com
Website: www.astralint.com

Printed at : **Replika Press Pvt. Ltd.**

FROM THE DESK OF THE DIRECTOR

It is with great pride and satisfaction that I present the first volume of the Plant Tissue Culture book, a meticulously crafted and timely compendium developed by the ICAR-Central Institute of Temperate Horticulture (ICAR-CITH), Srinagar. This book embodies our unwavering commitment to advancing biotechnological interventions for promoting sustainable, high-value horticulture in temperate agro-climatic zones.

Plant tissue culture has emerged as a transformative tool in contemporary plant science-enabling precision-driven crop improvement, rapid clonal propagation of disease-free planting material, and long-term conservation of superior germplasm. Acknowledging the far-reaching implications of this technology, ICAR-CITH has strategically integrated advanced tissue culture methodologies into its core research programmes, capacity-building initiatives, and on-ground applications, especially in the context of temperate fruit and vegetable crops that are integral to the regional agroeconomy.

This book is the outcome of concerted efforts by a team of experienced scientists and skilled researchers. It offers a comprehensive exploration of theoretical foundations, optimized protocols, and hands-on guidance on a spectrum of in vitro techniques including callus induction, micropropagation, anther culture, somatic embryogenesis, and genetic transformation. The content has been carefully curated to cater to a diverse audience ranging from undergraduate and postgraduate students to researchers, academicians, and industry professionals engaged in plant biotechnology and horticulture.

It is my sincere hope that this book not only serves as a robust laboratory reference but also acts as a catalyst for fostering innovation, knowledge dissemination, and interdisciplinary collaboration. In an era where climate-resilient, resource-efficient horticultural systems are indispensable, such knowledge resources will be instrumental in shaping future-ready scientists, practitioners, and policy influencers.

I extend my warm wishes to all users of this book for a productive, insightful, and enriching experience.

All the best

Dr. M. K. Verma
Director,
ICAR-Central Institute of Temperate Horticulture
Srinagar, Jammu and Kashmir, India

August, 2025

PREFACE

Plant tissue culture stands at the frontier of plant science, offering transformative tools for the propagation, conservation, and genetic enhancement of plant species. In the context of increasing demands for food security, clean planting material, and climate resilience, tissue culture has emerged as an indispensable platform in both academic research and commercial horticulture. With its growing relevance, the need for a technically sound, regionally tailored, and scientifically robust resource has never been more pressing especially for temperate horticultural crops, which are often underserved in mainstream literature. This book is a direct response to that need.

Compiled by a multidisciplinary team of researchers and academicians at the ICAR-Central Institute of Temperate Horticulture (ICAR-CITH), Srinagar, this volume reflects our collective experience and deep engagement with the theory and practice of plant tissue culture. It covers both foundational principles and specialized methodologies, offering comprehensive protocols suited for temperate crops such as apple, walnut, cherry, and others. The content has been structured to serve a dual purpose: to guide early-career researchers and students through essential laboratory procedures and to provide advanced knowledge on in vitro systems including haploid production, soma clonal variation, micrografting, and genetic transformation.

In addition to standard practices, the book ventures into emerging domains such as nanotechnology integration, biosensor applications, and biotization. These chapters are designed to spark innovation and encourage interdisciplinary exploration in plant biotechnology. Special focus has been given to reproducibility, scalability, and clarity of protocols, so that readers, regardless of laboratory resources can implement these techniques with confidence.

We have also included dedicated sections on infrastructure planning, contamination management, and troubleshooting strategies, recognizing that technical consistency and practical problem-solving are as critical as theoretical knowledge for achieving success in tissue culture laboratories. Each chapter is supplemented with insights drawn from real-world challenges and experimental optimization carried out at ICAR-CITH, making this book both scientifically rigorous and application-driven.

What distinguishes this volume is not just its technical depth, but its contextual relevance to temperate horticulture, an area of significant economic and ecological importance in India and globally. By bridging conventional and modern practices, this book aims to support the production of elite, disease-free planting material, promote germplasm conservation, and serve as a catalyst for biotechnological innovations in horticultural science.

It is our sincere hope that this book will serve as a comprehensive reference, a practical guide, and an inspirational resource for students, researchers, teachers, and professionals alike.

Dr. Javid Iqbal Mir

Dr. Sadiah Shafi

CONTENTS

CHAPTER 1

INTRODUCTION TO PLANT TISSUE CULTURE

Plant Tissue Culture

Plant tissue culture (PTC) is a basic and fundamental science within the field of plant biotechnology which aids in the understanding of plant growth and development at the cellular level. It is the most acquainted in vitro technique used for the propagation and study of plant tissues, cells and organs under aseptic and controlled environmental conditions. PTC is widely used for plant multiplication and propagation, on a larger scale on a nutrient medium of known concentration. It has huge applications in conservation, genetic modification and production of disease-free plants. Because to his theories on cellular totipotency in early 20th century, German scientist Gottlieb Haberlandt is regarded as the founder of plant tissue culture. Since then, noteworthy advancements have been made, making it an important tool in contemporary agriculture and horticulture. As Frederick Campion Steward famously stated, *"A single cell has within it all the information necessary to create an entirely new organism."* The statement emphasizes the totipotency principle, which describes a plant cell's capacity to proliferate into an entire plant. Growth regulators like auxins and cytokinin's, a nutrient-rich medium with vital macro- and micronutrients, aseptic conditions, and a regulated environment with ideal temperature, light, and humidity are all necessary for successful plant tissue culture.

This manual provides a step-by-step guide to essential plant tissue culture techniques, including the preparation of culture media, sterilization procedures, selection and preparation of explants, callus induction, organogenesis, somatic embryogenesis, micropropagation, and the hardening and acclimatization of tissue-cultured plants. As T. Murashige emphasized, *"Plant tissue culture is not only a tool for plant propagation but a means to explore plant biology at the cellular level, offering insights into genetics, development, and physiology."* By following the procedures outlined in this manual, students and researchers will gain practical knowledge and skills essential for mastering plant tissue culture and exploring its diverse applications in plant science.

Principles of Plant Tissue Culture

Plant tissue culture fundamentally aims to produce numerous plants genetically identical to a parent plant. This process utilizes an "explant," a small, dissected plant part, to develop a complete plant. The technique's effectiveness stems from the totipotency of most plant cells, meaning each cell contains the genetic information and cellular machinery necessary to generate an entire organism.

The basic principles of plant tissue culture are grounded in two key concepts:

1. **Cell plasticity:** Plants have developed a remarkable ability to adapt to extreme conditions due to their longevity and stationary nature. This adaptability enhances plant development and growth. When cultured in vitro, plant cells and tissues typically exhibit high plasticity, allowing one type of organ or tissue to initiate from another, ultimately leading to the regeneration of a whole plant.

2. **Totipotency**: This concept forms the foundation of successful plant tissue culture. The theory of totipotency posits that each cell can regenerate into a complete plant. Every somatic cell possesses the same genetic makeup (DNA sequence) as a zygote, thus having the potential to express all the organism's properties. Various plant parts, including single cells, protoplasts (plant cells without cell walls), and fragments of leaves, stems, or roots, can often generate a new plant when provided with appropriate nutrients and plant hormones in culture media. Due to the practical challenges of handling single cells, tissue culture typically begins with a tissue or organ from the plant. Consequently, plant tissue culture is often referred to as plant cell, tissue, and organ culture.

Importance of Plant Tissue Culture

- Plant tissue culture technology is extensively utilized for mass production of specific plant varieties. Beyond its research applications, plant tissue culture techniques have recently gained significant industrial importance in plant propagation, disease eradication, plant enhancement, and secondary metabolite production.
- Tiny tissue fragments (called explants) can be employed to generate hundreds or thousands of plants in an ongoing process. A single explant can be multiplied into several thousand plants within a relatively brief period and limited space under regulated conditions, regardless of seasonal or weather factors, throughout the year.
- Micropropagation has successfully cultivated and preserved endangered, threatened, and rare species due to its high multiplication rate and minimal requirements for initial plant numbers and space.

Explant Propagation

- Furthermore, plant tissue culture is regarded as the most effective technology for crop improvement through the generation of somaclonal and gametoclonal variants.
- Micropropagation technology offers immense potential for producing superior quality plants, isolating useful variants in well-adapted, high-yielding genotypes with enhanced disease resistance and stress tolerance capabilities.
- Moreover, plant tissue culture has garnered significant interest from molecular biologists, plant breeders, and industrialists, as it aids in enhancing economically important plants. Additionally, tissue culture substantially contributes to understanding the patterns and factors responsible for plant growth, metabolism, morphogenesis, and differentiation.

CHAPTER 2

PRECAUTIONS AND GUIDELINES FOR WORKING IN THE TISSUE CULTURE LABORATORY

The success of tissue culture techniques hinges not only on precise protocols but also on rigorous adherence to aseptic techniques and laboratory best practices. This chapter provides a comprehensive overview of the critical precautions required in a tissue culture laboratory to prevent contamination, ensure safety, and maintain experimental fidelity. Here are some basic guidelines which should be adhered:

1. Keeping Work Areas Sterile

- Clean all working surfaces with 70% alcohol or 1% bleach before starting any procedure. Repeat the cleaning at regular intervals.
- Turn on UV lights inside the airflow cabinets for 20 to 30 minutes before use. The hood must be unoccupied during this time.

- Run the airflow hood for at least 15-30 minutes before starting work.
- Always keep windows and doors closed while working to prevent contaminants from entering.
- Avoid placing large objects inside the airflow hood that might block the flow of clean air.
- Never bring unclean or contaminated materials inside the sterile work area.
- Tools such as scalpels, forceps, and spatulas must be sterilized by autoclaving and flame-sterilized between steps.

2. Personal Hygiene and Conduct

- Always wear a clean lab coat, gloves, a face mask, and a hair cover while in the lab.
- Use clean shoe covers when entering specially designated sterile rooms.
- Wash your hands thoroughly with antiseptic soap before entering the lab.

- Regularly disinfect your gloves with 70% alcohol, especially after touching anything outside the clean area.
- Avoid talking, coughing, or sneezing when close to cultures or open materials.
- Maintain focus and avoid unnecessary distractions during work.

3. Media Preparation and Handling

- Sterile media is critical for success. Autoclave culture media at 121 C for 15-20 minutes at 15 pounds of pressure.
- Heat-sensitive ingredients like plant hormones and vitamins must be sterilized using special filters and added after media cools.
- Label all media containers with the preparation date, batch number, and pH.
- Store prepared media in a cool, dark place to maintain its quality.

- Never warm media unnecessarily; allow it to reach room temperature naturally to preserve important nutrients like glutamine.

4. Explant Selection and Sterilization

- Choose healthy, young plant tissues like shoot tips or meristems for culturing.
- Use proper sterilizing solutions such as 1-2% bleach, 0.1% mercury chloride, or 70% alcohol.
- After sterilizing, rinse explants thoroughly with sterile distilled water to remove any harmful residues.

5. Handling Culture Vessels and Transfers

- Handle all vessels gently to prevent contamination or breakage.
- Perform all transfers only inside the airflow hood.
- Do not overcrowd the working area inside the hood to maintain proper air circulation.
- Use only sterilized tools and avoid touching the mouth of culture vessels.

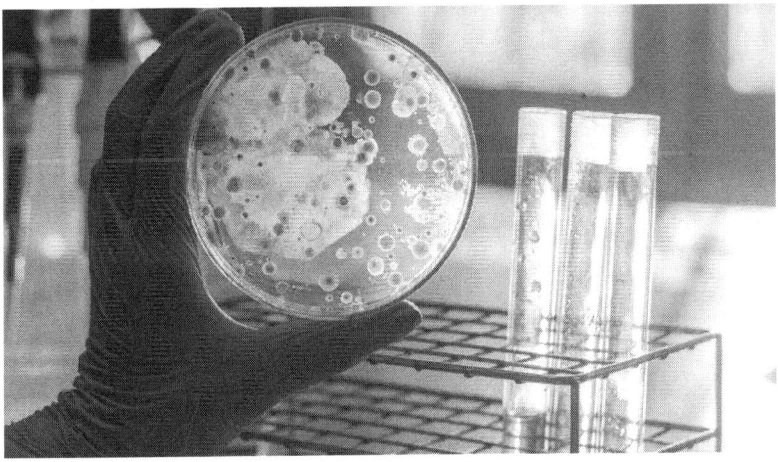

- Keep the incubation environment stable with correct light, temperature, and humidity according to the plant's needs.
- Avoid opening culture vessels unless necessary during incubation.
- Regularly check for any signs of contamination and remove any affected cultures immediately.

6. Equipment Maintenance and Calibration

- Ensure that equipment like autoclaves is regularly tested using indicators and that their performance is recorded.
- Calibrate pH meters and balances frequently using standard reference materials.

- Replace airflow cabinet filters as recommended and verify that airflow is functioning properly.
- Shut off equipment like centrifuges at the end of the day, and clean spills immediately with bleach, distilled water, and 70% alcohol.

7. Keeping the Laboratory Clean and Managing Waste

- Clean floors, corners, and benches regularly to reduce the buildup of dust and fungal spores.
- Clearly separate sterile work areas from non-sterile ones.
- Dispose of used media, contaminated cultures, and plastics into biohazard bins.
- Dispose of broken glass and sharp items into dedicated sharps containers.
- Always autoclave contaminated waste before disposal.
- Wash used media bottles with tap water and discard them in regular trash if safe.
- Discard contaminated flasks and plates after draining media into autoclave bags placed in grey buckets.

8. Rules for Using the Tissue Culture Room

- Never bring in materials that could carry contamination, and avoid carrying anything contaminated back into shared spaces.
- Keep laminar flow hoods clean and uncluttered at all times.
- Only place cuvettes inside hoods at night, not during active work hours.
- Always wipe down laminar hoods before and after use.
- Clean all spills first with distilled water to remove salts, then with 70% alcohol on surfaces including: Sliding glass sash, Vacuum knob, Pipette holders, Front grille, Armrests and chair backs
- After using the vacuum hose, bleach it, rinse with water, pull 70% alcohol through it, and wipe its exterior.
- Empty all pipette jars after use. Drain pipettes in bleach before transferring them to autoclave bags in grey buckets.
- Discard glass pipettes in sharps containers; discard plastic pipette tips into autoclave bags (not bleach jars).
- Do not put plastic flasks or boxes into sharps containers; only true sharps or contaminated items go there.
- The water bath is only for thawing cells or virus samples. Clean it immediately if contaminated.
- Shut off microscopes after use, wipe down spills, and properly clean haemocytometers and well plates used for cell counting.

9. Emergency Procedures

- For chemical spills, immediately use the correct spill kit and wear protective equipment.
- For biological spills, disinfect with 70% alcohol or bleach immediately.
- Handle open flames like Bunsen burners with care and turn them off after use.
- Never overload electrical sockets.
- Know where fire extinguishers, emergency exits, and first aid kits are located.
- Report any accidents or injuries immediately to your supervisor.

Important Note:

Every person working in the lab must strictly follow these rules.

CHAPTER 3

EQUIPMENT AND INSTRUMENTATION SPECIFICATIONS FOR PLANT TISSUE CULTURE LABORATORIES

To establish a well-equipped tissue culture laboratory by identifying and understanding the purpose of essential equipment and instruments critical for efficient and contamination-free plant tissue culture operations. Here are the core Equipment's and Instrumentation and their functional roles:

1. Electronic Analytical Balance

Accurate weighing is the cornerstone of reliable media preparation. A balance with 0.1 mg readability and a weighing capacity of 60–520 g ensures the precision necessary for preparing stock solutions and media formulations without deviation.

Key Function: Weighing fine quantities of plant growth regulators, nutrients, and chemicals with high accuracy.

2. pH Meter

Maintaining the correct pH is vital for nutrient uptake and enzymatic reactions during tissue culture. Modern pH meters use specialized glass electrodes and reference systems to translate the electrochemical potential difference into accurate pH readings, guided by the Nernst equation.

Key Function: Calibration and adjustment of media and solution pH to ideal physiological levels.

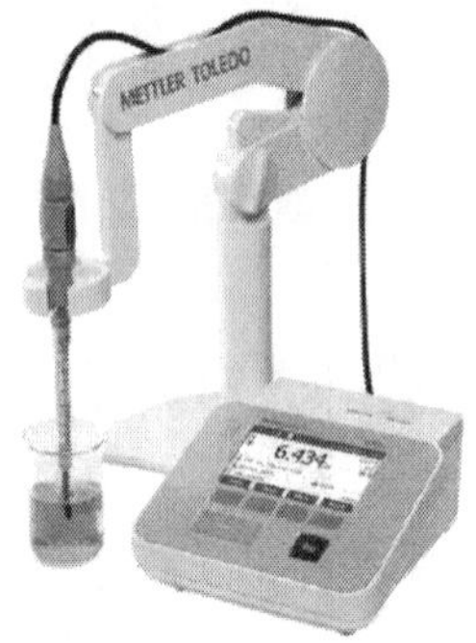

3. Autoclave

An autoclave is used to sterilize culture media, glassware, instruments, and other consumables by applying high-pressure saturated steam at 121 C and 15 psi.

Key Function: It ensures complete elimination of microbial contaminants to maintain aseptic conditions essential for tissue culture success.

4. Thermostatic Water Bath

Precise temperature control enhances the performance of various reagents and biological samples. Water baths maintain samples at a constant temperature, using sensor-driven feedback systems for stability.

Key Function: Pre-warming media, maintaining enzyme activity, and heat-treating sensitive solutions.

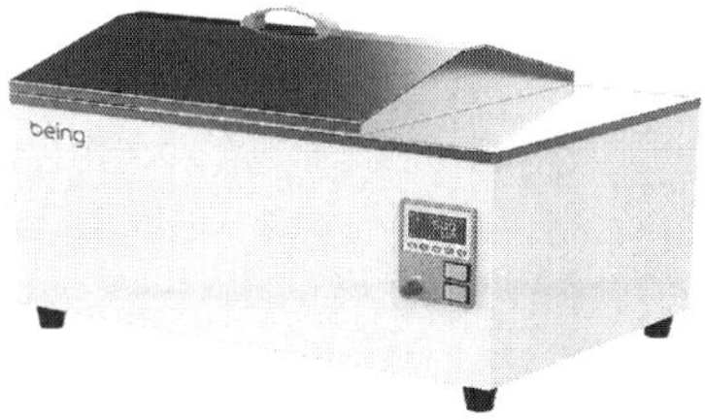

5. Hot Air Oven

When moisture-based sterilization is unsuitable, dry heat becomes essential. Hot air ovens sterilize heat-resistant materials by sustaining high temperatures uniformly across the chamber.

Key Function: Dry sterilization of metallic instruments, glass Petri dishes, and culture vessels.

6. Microwave Oven

An auxiliary sterilization tool, microwave ovens utilize electromagnetic radiation to generate molecular heat, effectively decontaminating certain media components and reusable plastics.

Key Function: Rapid sterilization of specific materials and preliminary media treatment.

7. Water Purification System

Tissue culture demands the highest water purity to prevent ion-based toxicity or microbial contamination. Systems combining reverse osmosis, deionization, and ultra-filtration provide ultrapure water tailored for sensitive cultures.

Key Function: Preparation of sterile and contaminant-free culture media.

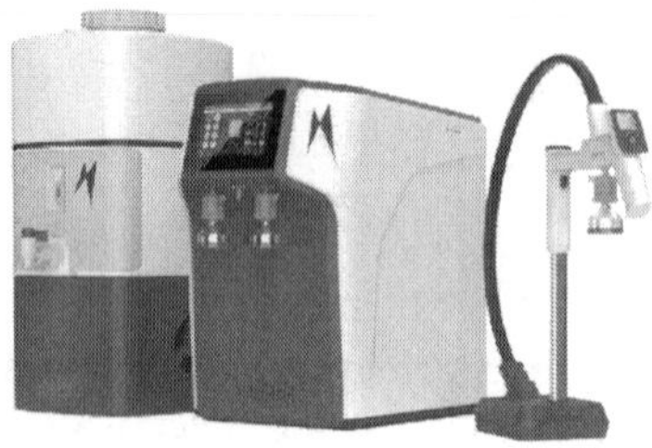

8. Laminar Air Flow Cabinet

An indispensable unit for aseptic manipulations, laminar flow cabinets rely on HEPA filters to maintain a sterile air curtain across the workspace.

Key Function: Prevention of airborne contamination during explant inoculation and subculturing processes.

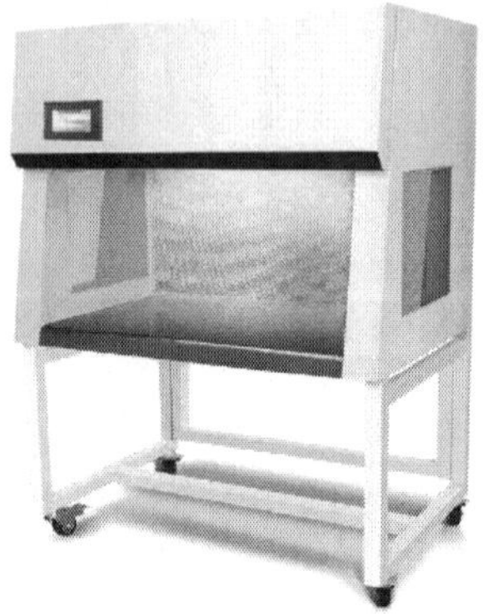

9. Vortex Mixer

Homogeneous mixing of reagents, suspensions, and samples is critical for reproducibility in tissue culture procedures. The vortex mixer imparts rapid orbital motion, ensuring efficient solubilization or resuspension.

Key Function: Quick and uniform mixing of culture media, reagents, and cellular suspensions.

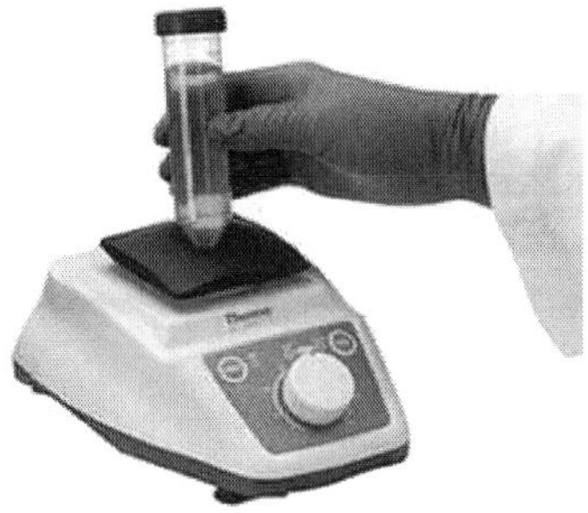

10. Air Conditioning System

Controlled temperature and humidity are essential environmental parameters that impact tissue viability and growth.

Key Function: Regulation of the laboratory's ambient environment to prevent stress-induced tissue responses.

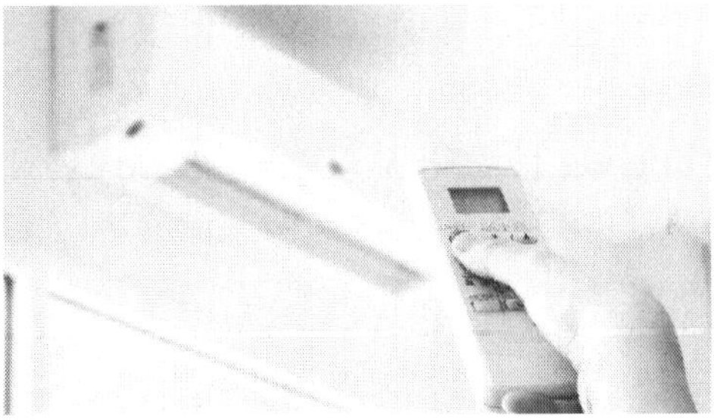

11. Centrifuge

Centrifugation separates cellular components or concentrates suspensions using centrifugal force, critical for downstream applications in plant tissue handling.

Key Function: Harvesting cells, isolating subcellular fractions, and clarifying media solutions.

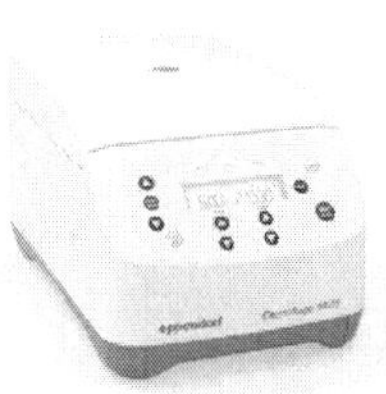

12. Photoperiodic Growth Racks (Castor Racks)

Simulating natural light-dark cycles is essential for morphogenesis and photo responsive growth in plant cultures.

Photoperiodic racks equipped with PAR lighting systems and programmable timers ensure consistent environmental stimulation.

Key Function: Providing controlled light exposure to support photosynthesis and plantlet development.

13. Refrigerator

Proper storage of volatile and temperature-sensitive compounds extends their functional lifespan and maintains consistency in experimental protocols.

Key Function: Preservation of stock solutions, culture additives, and plant hormones at 2–8 C.

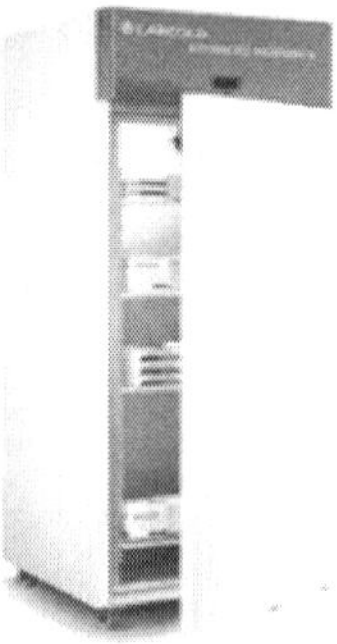

14. Freezer

Ultra-low temperature storage is often required for enzymes, specialized reagents, and unstable media components.

Key Function: Long-term preservation of delicate biological materials, ensuring experimental continuity.

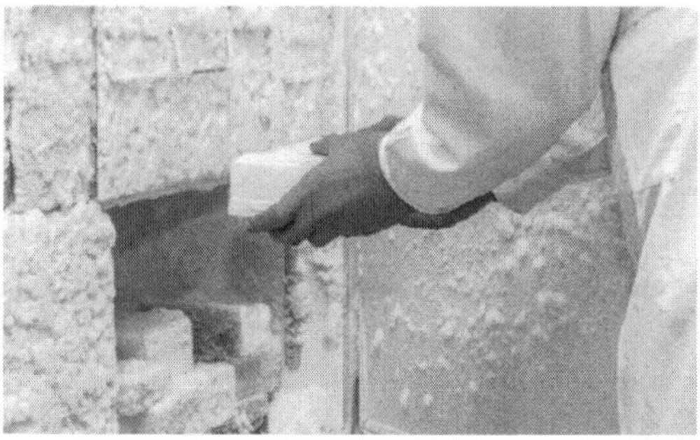

15. Hot Plate with Magnetic Stirrer

A combined heating and stirring device are used to dissolve media components and homogenize solutions during preparation.

Key Function: It ensures even heating and thorough mixing of culture media, essential for consistency and quality.

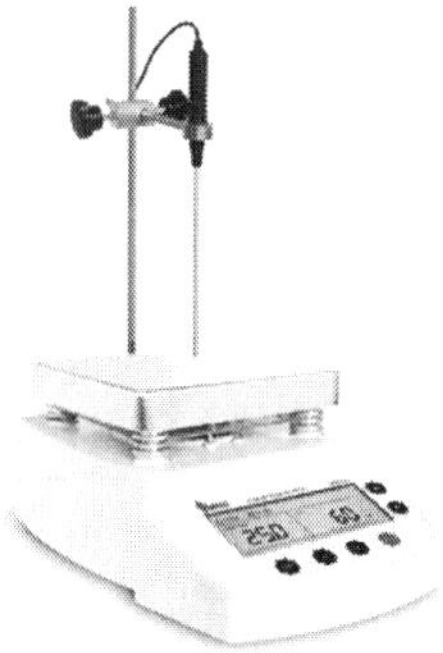

16. Incubators / Growth Chambers

These chambers control temperature, light, and humidity to create optimal growth conditions for cultured tissues.

Key Function: They simulate natural environmental conditions to support cell division, differentiation, and plantlet development.

17. Microscopes

Stereo or binocular microscopes allow for magnified viewing of explants and cultures during dissection and observation.

Key Function: They aid in precise manipulation of plant tissues and early detection of contamination or growth anomalies.

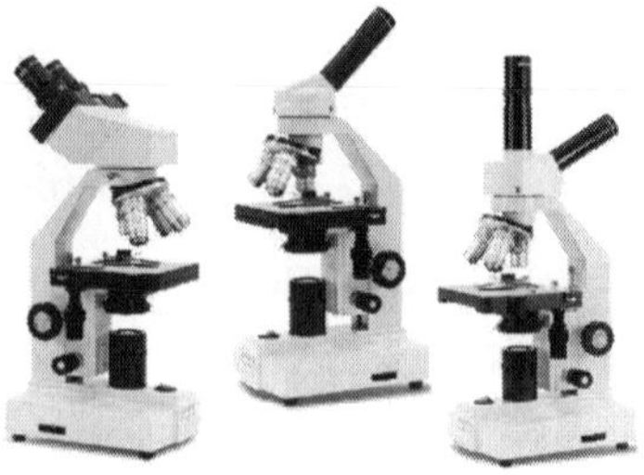

18. Humidity Control System

A dedicated humidifier or integrated humidity control in growth chambers maintains optimal relative humidity (40%–80%).

Key Function: It prevents dehydration of cultures and maintains turgor pressure critical for tissue survival.

19. Fume Hood

A fume hood provides safe ventilation for handling volatile chemicals during media preparation.

Key Function: It protects personnel from inhaling harmful vapours and maintains a safe working environment.

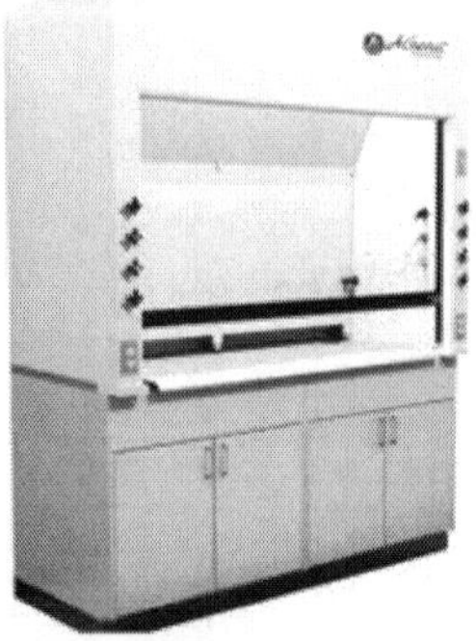

20. Fire Extinguisher

An ABC type fire extinguisher is installed as a precautionary measure.

Key Function: It provides immediate response capabilities in case of fire emergencies.

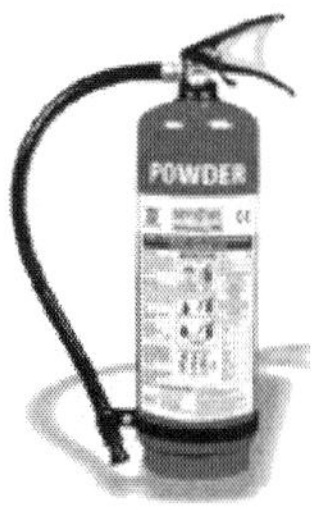

21. Dehumidifier

A dehumidifier may be used if environmental humidity exceeds optimal levels.

Key Function: It protects sensitive cultures from excessive moisture, which could promote microbial growth.

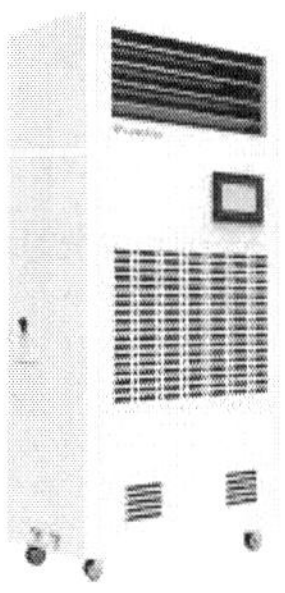

CHAPTER 4

INFRASTRUCTURE REQUIREMENTS FOR PLANT TISSUE CULTURE LABORATORIES

Establishing a high-performing tissue culture laboratory demands meticulous planning of infrastructure, utilities, and operational spaces. Each structural element must promote sterility, efficiency, and environmental control to ensure optimal plant tissue development and minimize the risk of contamination.

Structural Design and Utilities

- **For the success of tissue culture operations, laboratories must be strategically located away from sources of pollution.**
- Labs should also be located away from harvesting fields to minimize seasonal contamination.
- Growth and transfer rooms should be well-insulated, such as with double-wall construction, to conserve energy.
- Tissue culture facilities need ample high-quality water and wastewater disposal.
- Backup power is essential for transfer, growth rooms, and critical equipment to prevent disruptions from power outages.
- Single-level structures with efficient layouts, pass-throughs, and minimal backtracking are preferred for easy material flow and maximum cleanliness.
- Strict aseptic practices, including double-door hatches for material transfer, are essential to prevent microbial contamination.

Media Preparation Room

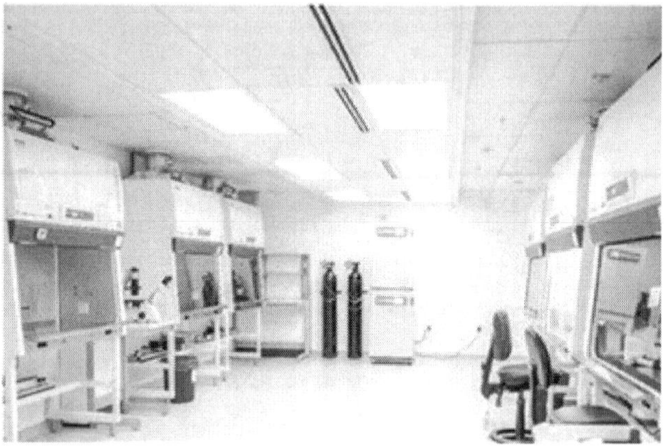

Media Preparation Room

The media room serves as the backbone of culture success, requiring high-purity water sourced from distillation, reverse osmosis, or Milli-Q systems. The preparation area should be ergonomically organized, equipped with:

- Workbenches at standing height for ease during long preparation sessions.
- A deep freezer for long-term storage of stock solutions, enzyme preparations, and sensitive organic additives like coconut milk.
- A refrigerator for short-term storage of reagents, chemicals, and plant material.
- Plastic carboys dedicated to storing purified water.
- High-precision analytical balances for accurate weighing of media components.
- Hot plate-cum-magnetic stirrers to dissolve media constituents efficiently.
- Calibrated pH meters for adjusting the pH of culture media.
- Pipettes and sterile pipette tips for precise reagent handling.
- Aspirators or vacuum pumps for facilitating membrane-based filter sterilization.
- Steamers for rapidly melting solidified agar.
- Autoclaves or, where applicable, domestic pressure cookers for media sterilization.
- Refrigerators and freezers can be located nearby, but weighing should ideally be done elsewhere or in a designated dry area within the media room.

Culture Vessels

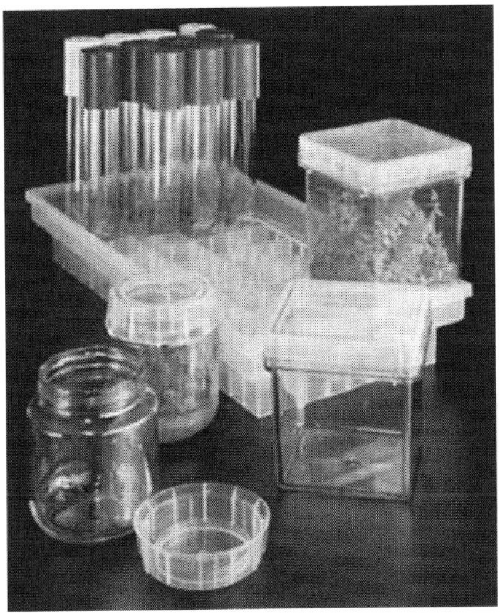

Culture Vessels

- Various vessels are used for plant tissue culture, often based on experiment needs or user preference. Standard work often uses glass test tubes, while micropropagation may utilize wide-mouth or milk bottles. Only borosilicate/Pyrex glassware should be used, as soda glass can be toxic.
- Plasticware, including autoclavable and disposable pre-sterilized vials, is increasingly replacing glassware in many labs, especially for specialized culture work.
- Various plastic and metal cap closures exist, including transparent, autoclavable polypropylene caps with membranes. Closure choice depends on availability and cost, but it's crucial to ensure it doesn't hinder plant growth.

Growth Room

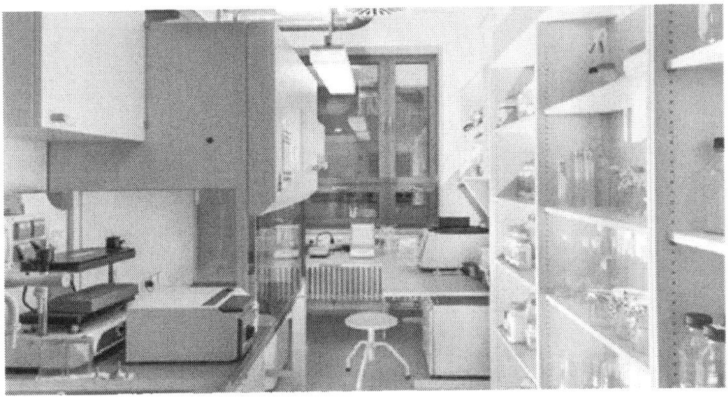

Growth Room

- Culture incubation rooms require precise temperature control (typically 25±2 C) achieved with air conditioners, heaters, and a temperature controller.
- Specialized incubators with built-in lights can provide higher or lower temperatures and can be placed outside the main culture room.
- Incubators placed outside the culture room must be protected from tampering. Multiple growth rooms can offer varied conditions. Cleanliness is crucial to prevent contamination.
- Walls and flooring should be easy to clean. Rounded wall junctions minimize cobwebs. Cultures typically grow in diffuse light (<1 klx), but provisions for higher light (5-10 klx) and darkness are also needed.
- Low humidity (<50%) requires humidification to prevent media drying, while high humidity increases contamination risk due to damp cotton plugs.

CHAPTER 5

GUIDE TO PREPARING SOLUTIONS AND PERFORMING CALCULATIONS

Solutions form the foundation of plant tissue culture media. Accurate preparation of stock solutions ensures consistency, reproducibility, and the success of in vitro culture. This chapter outlines the preparation of essential solutions, including macronutrients, micronutrients, vitamins, plant growth regulators, carbon sources, gelling agents, and sterilization protocols.

1. Molar Solutions (M)

A molar solution contains the gram molecular weight (molecular weight in grams) of a substance dissolved in water, with the final volume made up to one litre. Example:

1 M NaCl = 58.5 g of NaCl dissolved and volume adjusted to 1 L with distilled water.

2. Molal Solutions (m)

A molal solution is prepared by dissolving the gram molecular weight of a substance into one litre of solvent (without adjusting the volume after solute addition).

3. Normal Solutions (N)

Normal solutions involve dissolving the gram equivalent weight of a substance in water and making the final volume up to one litre.

Equivalent weight = Molecular weight (number of substituted H^+ ions or valency)

Example: 1 N H_2SO_4 = 49 g H_2SO_4 per litre (because H_2SO_4 can donate two H^+ ions).

4. Percent (%) Solutions

A percent solution is the number of grams of solute in 100 mL of solution. Example:

10% NaCl solution = 10 g NaCl in 100 mL total volume.

5. ppm Solutions (Parts Per Million)

ppm solutions refer to milligrams of solute per litre of solution. Example:

1 ppm = 1 mg of chemical dissolved in 1 L of solvent.

6. Millimolar (mM), Milli molal, and Milli normal Solutions

When milligrams (instead of grams) are used for solution preparation, the resulting solutions are referred to as millimolar (mM), milli molal, or milli normal.

Units Used in Solution Preparation:

mg/L (milligrams per litre):

- 10^{-6} = 1.0 mg/L or 1 ppm
- 10^{-7} = 0.1 mg/L
- 10^{-8} = 0.001 mg/L or 1 µg/L

Molarity (M):

- 1 M = Molecular weight in g/L
- 1 mM = Molecular weight in mg/L (10^{-} M)
- 1 µM = Molecular weight in µg/L (10^{-6} M)

Example Calculation for Conversion:

- Auxin (2,4-D) molecular weight = 221.0 g/mol

  1 M solution = 221.0 g/L

1 mM solution = 221.0 mg/L

1 µM solution = 0.221 mg/L

Converting mg/L to mM:

- Example for $CaCl_2$ $2H_2O$ (Molecular weight = 147.018 g/mol)

= 40.08 + 2 x 35.453 + 4 x 1.008 + 2 x 16 = 147.018

(the atomic weights of Ca, Cl, H and O being 40.08, 35.453, 1.008 and 16.0 respectively).

If, 440 mg/L of $CaCl_2$ - $2H_2O$ is to be converted into mM; then

The number of mM $CaCl_2$. $2H_2O$ = No. of mg $CaCl_2$ - $2H_2O$/ Molecular weight of $CaCl_2$. $2H_2O$ = 440/ 147.019 = 2.99 mM

Thus, 440 mg/ L $CaCl_2$ - $2H_2O$ = 2.99 mM

Preparation of Solutions

Preparation For solids:

Chemical	1 Molar	1 Molal	1 Normal	10 per cent	1000 ppm
HCl	36.5-gram HCl in per litre solution	36.5-gram HCl was added in 1 litre of distilled water.	36.5-gram HCl in per litre solution.	10 ml HCl in 100 ml solution	1 ml HCl in per litre solution.
H2SO4	98-gram H_2SO_4 in per litre solution.	98-gram H_2SO_4 was added in 1 litre of distilled water.	49-gram H_2SO4 in per litre solution.	10 ml H_2SO_4 in 100 ml solution.	1 ml H_2SO_4 in per litre solution.
NaCl	58.5 grams NaCl in per litre solution	58.5 grams NaCl were added in 1 litre of distilled water	58.5 grams NaCl in per litre solution.	10 grams NaCl in 100 ml solution.	1 grams NaCl in per litre solution.
$CaCl_2$	110 grams $CaCl_2$ in one litre solution	110 grams $CaCl_2$ were added in one litre of distilled water.	55 grams $CaCl_2$ in per litre solution.	10 grams $CaCl_2$ in 100 ml solution.	1 grams CaCl2 in per litre solution.

Preparation for liquid solutions:

For 1M HCl Solution,

If, HCl concentration is 37%, specific density = 1.19, molarity = 36.5

Then, volume of chemical (V)=Molecular weight (M) x 100 /Density x Purity

= 36.5 x 100/1.19 x 37

= 82.89 ml

For 1M H_2SO_4 Solution

If, H_2SO_4 concentration = 98%, specific density = 1.83 g/mL, molarity = 98

Then, volume of chemical (V)=M x 100 /Density x Purity

=98 x 100 /1.83 x 98

= 54.64 ml

Thus, 54.64 ml H_2SO_4 dissolved in distilled water and make the volume 1 L

Types of Solutions

1. Macronutrient Stock Solutions

Macronutrients (N, P, K, Ca, Mg, and S) are required in larger quantities. Common formulations include Murashige and Skoog (MS), Gamborg's B5, and others.

Preparation:

Prepare individual concentrated stocks (e.g., 10X or 20X) for each macronutrient. Store at 4 C and protect from light if necessary (e.g., phosphates).

Compound	Amount for 1 L of 10X Stock	Remarks
NH_4NO_3	16.5 g	Store in dark
KNO_3	19.0 g	
$CaCl_2$ $2H_2O$	4.4 g	
$MgSO_4$ $7H_2O$	3.7 g	
KH_2PO_4	1.7 g	Sensitive to light

2. Micronutrient Stock Solutions

Micronutrients (Fe, Mn, Zn, Cu, B, Mo) are required in trace amounts.

Preparation:

Prepare a combined 100X or 200X stock solution. Use chelated iron (Fe-EDTA) to avoid precipitation.

Compound	Amount for 1 L of 100X Stock	Remarks
H_3BO_3	6.2 g	
$MnSO_4$ $4H_2O$	16.9 g	
$ZnSO_4$ $7H_2O$	8.6 g	
KI	0.83 g	
Na_2MoO_4 $2H_2O$	0.25 g	Prepare fresh if possible
$CuSO_4$ $5H_2O$	0.025 g	
$CoCl_2$ $6H_2O$	0.025 g	
$FeSO_4$ $7H_2O$ + Na_2EDTA	27.8 mg + 37.3 mg	Chelate before combining

3. Vitamin Stock Solutions

Vitamins are essential for enzymatic and metabolic functions.

Preparation:

Prepare a 100X vitamin stock solution. Filter-sterilize (do not autoclave) and store at –20 C.

Compound	Amount for 1 L of 100X Stock	Remarks
Thiamine-HCl (Vitamin B1)	10 mg	Light-sensitive
Pyridoxine-HCl (Vitamin B6)	1 mg	
Nicotinic acid	1 mg	
Glycine	2 mg	Optional
Myo-inositol	100 mg	Usually added separately

4. Plant Growth Regulator (PGR) Solutions

Plant hormones like auxins, cytokinins, gibberellins, and abscisic acid regulate morphogenesis.

Preparation:

Dissolve PGRs in specific solvents (e.g., auxins in ethanol or NaOH; cytokinin's in HCl). Prepare concentrated stocks (1 mg/mL or 10 mM). Filter sterilize and store aliquots at –20 C or –80 C.

Hormone Type	Example	Solvent Required
Auxin	IAA	1 N NaOH or ethanol
Cytokinin	BAP	1 N HCl
Gibberellin	GA_3	Ethanol
ABA	ABA	Organic solvent + water

5. Carbon Source

Sucrose (2 to 5%) is the most common energy source.

Preparation:

Add sucrose directly into the media during preparation. Autoclave with the media unless otherwise specified.

6. Gelling Agents

Agar, phytagel, and gellan gum are used for solid media.

Preparation:

Weigh accurately (usually 0.7 to 1.0% w/v). Add before autoclaving. Dissolve completely by heating.

Precautions During Solution Preparation:

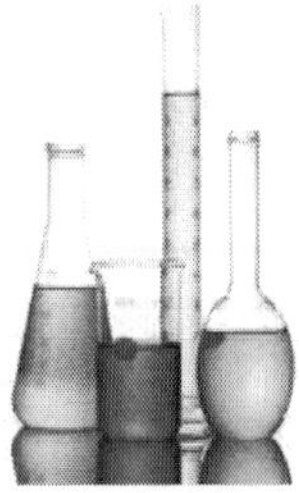

- Always use analytical-grade chemicals.
- Use freshly prepared or appropriately stored stock solutions.
- Minimize exposure to light for light-sensitive compounds.
- Ensure accurate pipetting and weighing.
- Maintain aseptic conditions to prevent contamination.

CHAPTER 6

MEDIA COMPOSITION AND PREPARATION FOR PLANT TISSUE CULTURE

Aim

To prepare Murashige and Skoog (MS) basal medium and accurately formulate and incorporate plant growth regulators (PGRs) for use in various stages of plant tissue culture.

Test tubes containing MS media

Principle

The MS (Murashige and Skoog, 1962) basal medium, formulated by T. Murashige and F. Skoog, is the most widely used nutrient medium for plant tissue culture and regeneration. It provides a complete spectrum of essential nutrients necessary for the growth, maintenance, and differentiation of plant explants or subcultures. The tissue culture medium typically consists of: 95% sterile water, Macronutrients and micronutrients (inorganic salts), Vitamins and amino acids (organic cofactors), Carbon source (commonly sucrose), Plant Growth Regulators (PGRs) (in micro-molar concentrations), Macronutrients are supplied in millimolar (mM) quantities, whereas micronutrients are present in micromolar (μM) concentrations. Vitamins function as enzymatic cofactors essential for key metabolic activities. A carbon source is indispensable because, under in vitro conditions, plant tissues often lack sufficient photosynthetic capacity, especially during initial culture stages. Plant growth regulators such as auxins and cytokinins are pivotal in regulating morphogenesis (shoot and root development) at concentrations typically ranging from 0.1 to 100 μM. The balance between auxins and cytokinins critically influences organogenesis: A high auxin-to-cytokinin ratio promotes root formation, A high cytokinin-to-auxin ratio favours shoot induction. Other growth regulators (e.g., gibberellins, abscisic acid) may be incorporated depending on the specific tissue culture system or experimental objectives.

Materials Required

Amber storage bottles; Plastic beakers (100 mL, 500 mL, 1000 mL); Measuring cylinders (500 mL capacity); Glass beakers (50 mL); Disposable syringes (5 mL); Disposable syringe filters (0.22 μm pore size); Autoclaved Eppendorf tubes (2 mL); Eppendorf tube stand; Plant Growth Regulators like Benzyl Aminopurine (BAP)-Cytokinin & Naphthalene Acetic Acid (NAA) – Auxin

Preparation of Stock Solution

Usually, the stock solution of macronutrients is prepared as 20x strength while a stock solution of all the micronutrients with 200x is generally prepared. Nutrient salts and vitamins are prepared as stock solutions as specified in table provided below and stored at 4 C to avoid frequent repetition during the optimization of the media and continuous culture. The required number of concentrated stocks is mixed to prepare 1 L of medium. The quantity of reach component of the salts and vitamins, required to make the stock solution is listed in the table given below (By the courtesy of Murashige T & Skoog F (1962), Plant 15: 473-497).

MS major salts	mg/L medium	500 ml stock (20X)
NH_4NO_3	1650 mg	16.5 gm
KNO_3	1900 mg	19 gm
$CaCl_2.2H_2O$	440 mg	4.4 gm
$MgSO_4.7H_2O$	370 mg	3.7 gm
KH_2PO_4	170 mg	1.7 gm

MS minor salts	mg/L medium	500 ml stock (200X)
H_3BO_3	6.2 mg	620 mg
$MnSO_4.4H_2O$	22.3 mg	2230 mg
$ZnSO_{4.}4H_2O$	08.6 mg	860 mg
KI	0.83 mg	83 mg
$Na_2MoO_{4.}2H_2O$	0.25 mg	25 mg
$COCl_{2.}6H_2O$	0.025 mg	2.5 mg
CuSO4.5H2O	0.025 mg	2.5 mg
MS Vitamins	**mg/ L medium**	**500 ml stock (200X)**
Thiamine (HCl)	0.1 mg	10 mg
Niacine	0.5 mg	50 mg
Glycine	2.0 mg	200 mg
Pyrodoxine (HCl)	0.5 mg	50 mg

Iron, 500ml Stock (200X)

Dissolve 3.725gm of Na_2EDTA (Ethylene diamine tetra acetic acid, disodium salt) in 250 ml distilled water. Dissolve 2.785 gm of $FeSO_4.7H_2O$ in 250 ml of water, Heat Na_2EDTA solution and add $FeSO_4$ solution gently by stirring.

Preparation of PGRs Stock Solution

The heat-labile plant growth regulators are filtered through the bacteria-proof membrane (0.22 μm) filter and added to the autoclaved medium after being cooled enough (less than 60 °C). The stocks (1 mM) of plant growth regulators are prepared based on the table given below.

Plant Growth Regulator	Nature	Mol. Wt.	Solvent
Benzyl amino purine	Autoclavable	225.2	1N NaOH
Napthalene acetic acid	Heat labile	186.2	Ethanol

The required amount of plant growth regulators is dissolved as listed in the table and the volume is raised by adding double distilled water or Milli Q water. The solutions are passed through disposable syringe filter (0.22 μm) and stocks are stored at –20 °C Similarly, the stock solution of other PGRs may be prepared as per the molecular weight and the volume required.

Preparation of MS media (1X strength)

1. In a sequence, different components are added into a beaker according to the list: Major and Minor nutrients, iron-EDTA, vitamins, myo-inositol, growth regulators (if thermostable and autoclavable), organic supplements, sucrose etc., by using the correctly sized graduated cylinders or pipettes or balance.

2. Water is added to just below the final volume (e.g. 800 ml volume for one litre medium)
3. pH of the medium is adjusted to the required value (e.g. pH 5.8 for MS) by adding dropwise while stirring 1N KOH or 1N HCl.
4. Required quantity of agar or any other gelling agent is added while the medium is being stirred.
5. The solution to brought to the final volume, i.e. 1 Litre and heated with continuous stirring until all the agar is dissolved and the solution becomes transparent.
6. The medium is dispensed in glass or polypropylene vessels and plugged with cotton plugs.
7. Culture medium is sterilized in an autoclave for 20 min at 121°C at 15 psi (105 kPa).
8. If the medium contains heat-labile substances:
    a. Steps 1-5 are followed except for the addition of heat labile substances.
    b. Culture medium is sterilized as such in a big Erlenmeyer flask without dispensing in vessels in an autoclave for 20-25 min at 121°C at 15 psi (105 kPa).
    c. The thermolabile compound solutions are filter sterilized using Millipore or any other filter assembly using 0.22 µm filter.
    d. After autoclaving, the medium is kept in a laminar airflow hood and allowed to cool to a temperature of around 50°C. The requisite quantity of the compound is added to the medium with the help of micropipettes while the medium is being stirred.
    e. The medium is dispensed into sterile containers (generally sterile petri dishes) under the hood of laminar airflow, provided the neck of the Erlenmeyer flask is passed over a flame before the medium is poured from it.
    f. Medium is allowed to cool and solidify in a laminar airflow hood.

CHAPTER 7

STERILIZATION TECHNIQUES IN PLANT TISSUE CULTURE

Aim

To study and understand various sterilization techniques used in plant tissue culture to ensure aseptic conditions and prevent microbial contamination of media, explants, and equipment.

Principle

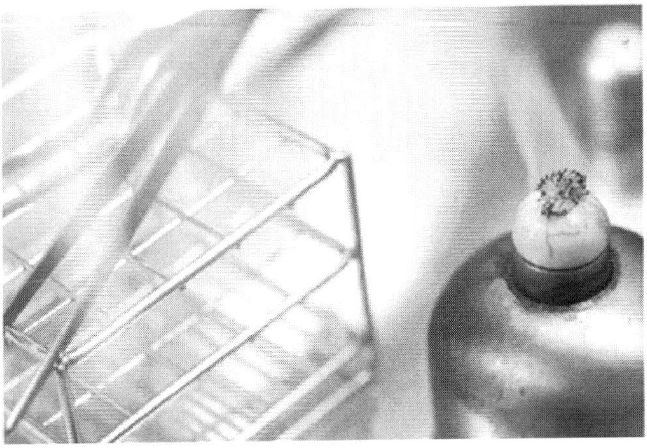

Aseptic conditions are fundamental for all manipulations in plant tissue culture. To maintain sterility, culture vessels, nutrient media, manipulation tools, and plant explants must be rendered free of microorganisms, including bacteria, fungi, viruses, and their spores. Sterilization is an absolute process that results in the complete destruction or removal of all forms of microbial life. In industrial and research contexts, sterilization is often defined operationally by achieving a Sterility Assurance Level (SAL) of 10^{-6}, meaning there is less than one chance in a million of a viable microorganism surviving post-treatment. Sterilization in plant tissue culture can be achieved through various methods such as heat (moist and dry), chemical agents, filtration, radiation, and aseptic workspaces like laminar airflow hoods. Infections during tissue culture operations may originate from: Airborne spores (fungal and bacterial); Contaminated explants carrying pathogens on their surfaces; The operator during handling.

Sterilization Procedures

The following sterilization strategies are routinely adopted to minimize contamination:

- Culture media, vessels, and instruments are sterilized by autoclaving at 121°C under 15 psi pressure for at least 20 minutes. Time and conditions may vary depending on the nature and volume of the material.
- Heat-sensitive compounds, such as stock solutions of plant growth regulators (PGRs), are sterilized by membrane filtration using 0.22 µm filters.
- Plant explants are surface sterilized using chemical disinfectants like sodium hypochlorite (NaOCl) or mercuric chloride ($HgCl_2$) before culture initiation.

Techniques of Sterilization

1. Physical methods:

a) Dry heat:

i) Hot air oven: Hot air oven is commonly used to sterilize glassware such as petri dishes. pipettes and assembled glass syringes. Dry heat at 160-180°C is applied for sterilizing materials that cannot be penetrated by steam and can tolerate high temperatures. Commonly used time and temperature for sterilization are:

1. 180°C for 20 minutes
2. 170°C for 40 minutes
3. 160°C for 60 minutes

Precautions:

1. The chamber door should not be immediately open after sterilization.
2. The chamber door should be opened when temperature has reached below 60°C otherwise glassware may get cracks.

ii) Incineration and Flaming: It is used to sterilize inoculating loops and wires by holding them almost vertically in Bunsen burner until red hot along their whole length almost to the tip of metal holder. Flaming is used to destroy microbes on the surface of scalpels, blades, forceps glass slides and mouths of culture flasks and tubes by slowly passing through Bunsen burner flame.

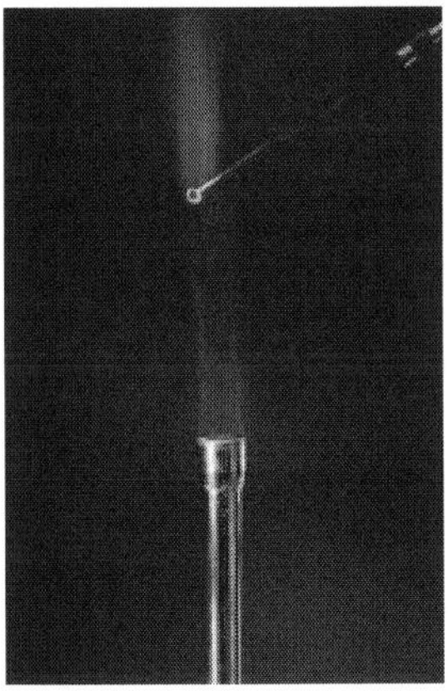

b) Moist heat sterilization

Microorganisms and their spores can be destroyed at lower temperatures by moist heat than dry heat.

Moist heat is used for sterilization of culture media and other liquids required for retaining their water content. This can be achieved by steam under pressure (Autoclaving), Tyndallization and Pasteurization).

i) Steam under pressure: Autoclaving provides steam under pressure at temperatures higher than 100°C. Steam developed in the autoclave kills microbes by denaturing their protein.

Generally, but not always Autoclave is operated at a pressure of 15 lb per sq. inch at a temperature of 121°C. The time of operation depends on the nature of material being sterilized.

Precautions:

1. Too much loading should be avoided because it prevents proper circulation of air.
2. All air in the chamber of autoclave must be completely replaced by saturated steam by keeping steam outlet open.

ii) Pasteurization below 100°C: This process is used to kill mesophillic vegetative bacteria in heat labile fluids by heating for 1 hour at 56°C or for 10 minutes at 65-75°C.

c) Radiations:

i) Ionizing Radiations: These are excellent sterilizing agents and penetrate deep into objects.

Gamma rays from cobalt-60 source are used for sterilization of antibiotics, hormones, plastic disposable supplies such as syringes. Gamma rays are also used to sterilize meat and other foods.

ii) Ultra-Violet Radiations: These radiations are emitted mostly from mercury vapour lamps emitting most of their radiations in the range of 250-260 nm and are bactericidal. But these radiations don't penetrate glass, dirt films. Commercial UV units are available for water treatment. UV lamps are sometimes places on ceilings of rooms, biological safety cabinets to sterilize the air and any exposed surface.

d) Filtration:

Filtration is used to sterilize solutions which are thermolabile liquids e.g. Serum, antibiotic solution, amino acid solution, vitamin solutions for removing bacteria physically.

Different types of filters are:

i) Syringe filters: Membranes of 13 mm and 25 mm diameter can be fitted in syringe like holders of stainless steel or polycarbonates. This is used to sterilize small volume of heat labile fluids.

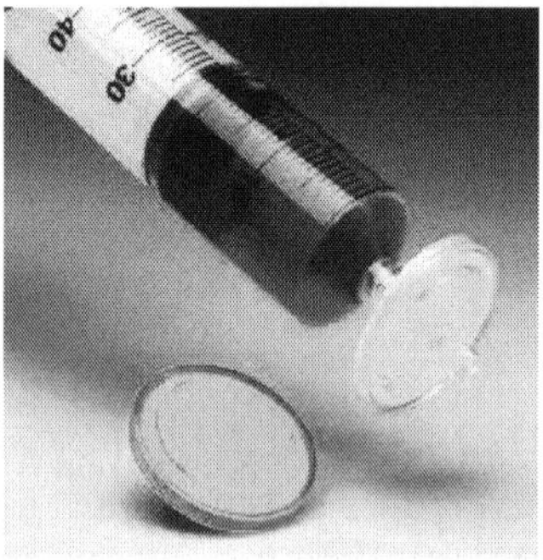

ii) Depth Filters: These filters consist of fibrous or granular materials that have been bounded into a thick layer filled with twisting channels of small diameter. The retention of microbes in solution is through entrapment and adsorption to the surface of the filter material. These filters are made up of different materials such as asbestos, diatomaceous earth in the Berkefeld filter, porcelain Chamberland, sintered glass filters.

iii) Membrane filters: Membrane filters have replaced depth filters and are made up from cellulose nitrate, cellulose diacetate, polycarbonates and polyesters. As these are less absorptive, they have faster rate of filtration for any given porosity. A variety of pores sizes are available, membranes with pores about 0.2 m are used for removal of vegetative cells but not viruses. The membrane filters can be sterilized by autoclaving. The filters are for a single use after sterilization.

iv) Air filters: large volumes of air may be freed from pathogens by passing through HEPA (high efficiency particulate air) filters. These employed to decontaminate the air in laminar air flow chamber.

2. Chemical methods:

a) Disinfection and antisepsis:

Disinfection is destruction of harmful microorganisms and chemicals used for this are called disinfectants. Antisepsis is disinfection of living tissue and chemicals used for this are called antiseptics. Term disinfectant generally used for chemical agents employed to disinfect inanimate objects. The disinfectant kills vegetative bacteria, fungi, viruses and occasionally spores by destruction of proteins, lipids or nucleic acids in the cell or its cytoplasmic membrane.

Rules for using disinfectants

1. The disinfectants are usually toxic and may damage or irritate the skin, conjunctiva and mucosa.
2. The use of gloves while handling the disinfectants is generally recommended.
3. These must always be used in correct concentration.

Common disinfectants used in laboratories:

1. Phenolic compounds: Lysol is an effective phenolic compound but is pungent and irritant to skin. Triclosan is a phenolic antiseptic used in antimicrobial soap. Hexylresorcinol used in throat instruments.
2. Halogens: A halogen is any of these five i.e. fluorine, chlorine, bromine, iodine and astatin
3. Aldehydes: Aldehydes like formaldehyde and glutardehyde denature microbial protein.
4. Alcohols: The most popular alcohols used in sterilization are ethanol (ethyl alcohol) and isopropanol (isopropyl alcohol). Alcohols are protein denaturants.
5. Heavy metals: Most of the heavy metals and heavy metal compounds or metallic salts have toxicity for microorganisms. The most toxic heavy metals are the mercury, silver and copper.

CHAPTER 8

PLANT TISSUE CULTURE TECHNIQUES

Aim

To understand and perform the sequential steps involved in plant tissue culture, including explant selection, inoculation, regeneration, and acclimatization under controlled in vitro conditions.

Principle

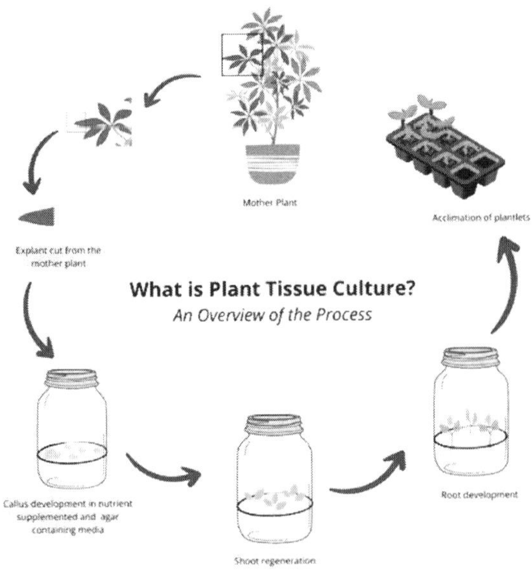

Plant tissue culture is the in vitro growing of parts of a plant such as organs, tissues or cells on an artificial nutritional medium. It is a combination of various techniques like micropropagation, callus culture, protoplast culture etc. and provide commercial products within limited space and time. The part of plant where from tissue culture is started is called explants. Now a day, micropropagation is commercialized for rapid propagation of horticultural plants. Since this culture is made in the laboratory, therefore it requires following steps.

1. Preparation of the media
2. Sterilization of the explants
3. Inoculation of the explants
4. Incubation of the explants in the culture room
5. Regular Subculture of the explants leading to regeneration of the plantlets
6. Acclimatization of the plantlets.

1. Preparation of the nutrient media:

(Refer Previous chapter), the media may be modified as per the type and source of explants and mode of regenerations.

2. Surface sterilization of the explants:

The explants may be opted any part of the plant such as shoot tip, apical buds, roots, leaf, endosperm, seed, cotyledons, embryo, male flower. The choices of explants depend upon the type and physiological condition of the explants. Whatever may be the explants it is essential to treat with the disinfectants e.g. sodium or calcium hypochlorite solution 0.3- 0.6%), mercuric chloride, alcohol followed by washing the explants with sterile distilled water. Other disinfectants and detergents may also be used as per the standard protocol of the particular plant from the particular explants. But the exposure and duration of the disinfectants is compromised with the vitality of the explants.

3. Inoculation of the explants:

Once surface sterilised, the explants is transferred onto the suitable nutrient medium (sterilized by autoclaving) in culture vessels under sterile conditions. Both steps surface sterilization and inoculation are carried out in the laminar flow hood to avoid the risk of contamination. The culture medium is also an excellent resource for the growth of contaminates such as fungus and bacteria, and the growth of the contaminates easily outpaces that of the cultured plant cells. Therefore, general procedures of microbiology are applied to avoid contamination. Basically, anything that may have the chance to make a direct contact with the culture medium and/ or explants has to be sterilized and manipulations should be designed such that there is no direct link between the culture and its nonsterile surroundings.

4. Incubation or growing the cultures in plant tissue culture room at optimum:

After inoculation the culture vessel is wrapped with the paraffin wax and allowed to incubate the explants in the culture room under optimum photoperiod, humidity and light intensity condition. The typical photoperiod required in most of the crop is 16 h day and eight hours night with temperature 25 C and relative humidity in the range of 50%-60%.

5. Regular Subculture of the explants leading to regeneration of the plantlets:

The explants grown on the media in the culture vessel consumes nutrients causing nutrient deficiency and due to growth of the explants there is scarcity of the space in the vessel as well. These necessities the need of the transfer in another culture vessel supplemented with the media. In addition, modification in the media components is required during different stages of the plant. Therefore, subculture is also required to meet the different media requirement during the regeneration into the plantlet.

6. Acclimatization of the plantlets:

The in vitro raised plantlets are in not in position to survive under natural condition directly. These plantlets have to be planted under the condition and the soil enriched with nutrients. Therefore in vitro raised plantlets are planted under greenhouse condition to make it hardened before planting into the natural environments.

Note:

This is the typical tissue culture techniques, media, the method of surface sterilization, and the subculture duration varies from species to species and the source and physiological condition of explants.

CHAPTER 9

ASEPTIC HANDLING OF EXPLANTS IN PLANT TISSUE CULTURE

Aim

Aseptic Inoculation and manipulation of various explants

Principle

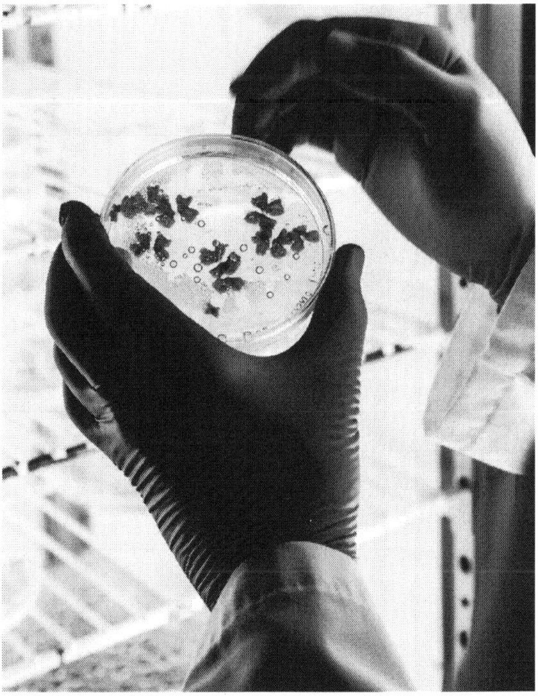

Aseptic technique is absolutely necessary for the successful establishment and maintenance of plant cell, tissue and organ cultures. The in vitro environment in which the plant material is grown is also ideal for the proliferation of microorganisms. In most cases the microorganisms outgrow the plant tissues, resulting in their death. Contamination can also spread from culture to culture. The purpose of aseptic technique is minimizing the possibility that microorganisms remain in or enter the cultures. This technique is absolutely necessary for the successful establishment and maintenance of plant cell, tissue and organ cultures. In most cases the microorganisms outgrow the plant tissues, resulting in their death. Contamination can also spread from culture to culture. The purpose of aseptic technique is to minimize the possibility that microorganisms remain in or enter the cultures. Explants require surface disinfestation before they can be placed in culture on the nutrient agar. This is generally accomplished by using diluted commercial chlorine bleach. Some explants, such as very small seeds or fern spores, are surface-disinfected in conical, capped centrifuge tubes and require centrifugation to pellet the seeds and decanting off the solutions with a pipette. Explants that float in the disinfectant can be wrapped in squares of cheesecloth to prevent their floating out of the beaker or test tube as the solutions are changed.

Contaminants in Plant Tissue Culture

A. Initial Contaminants

Most contamination is introduced with the explant because of inadequate sterilization or just very dirty material. It can be fungal or bacterial. This kind of contamination can be a very difficult problem when the plant explant material is harvested from the field or greenhouse. Initial contamination is obvious within a few days after cultures are initiated. Bacteria produce "ooze" on solid medium and turbidity in liquid cultures. Fungi look "furry" on solid medium and often accumulate in little balls in liquid medium.

B. Latent Contamination

This kind of contamination is usually bacterial and is often observed long after cultures are initiated. Apparently, the bacteria are present endogenously in the initial plant material and are not obviously pathogenic in situ. Once in vitro, however, they increase in titer and overrun the cultures. Latent contamination is particularly dangerous because it can easily be transferred among cultures.

C. Introduced Contamination

Contamination can also occur as a result of poor sterile technique or dirty lab conditions. This kind of contamination is largely preventable with proper care.

Detection of contaminants

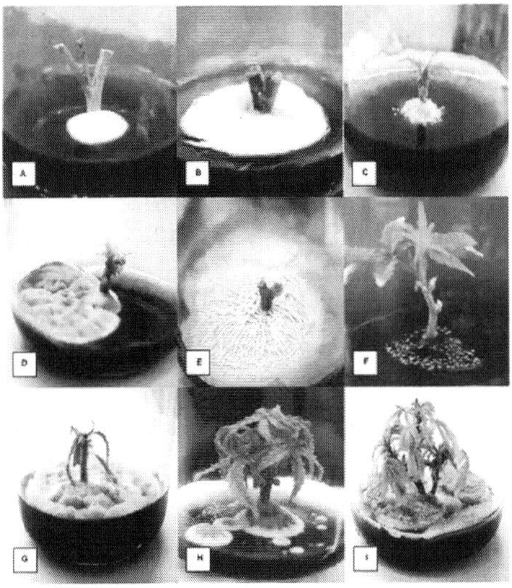

Contamination is usually detected by the "eyeball" method in research labs. However, indexing is possible, and is frequently done in commercial settings. This involves taking a part of the plant tissue and culturing it in media that are specific for bacteria and fungi. Media that have been used for this purpose include PDA (potato dextrose agar) and NB broth (with salts, yeast extract and glucose). This is the most reliable method for detecting bacteria and fungi, but, as indicated above, there may be infecting organisms that won't be detected.

Procedure of Aseptic Inoculation and Manipulation

A general procedure for preparing the explant is as follows:

1. Wash the explants in warm, soapy water and rinse in tap water. This procedure is very beneficial for stem, leaf, and shoot tip explants from the field or greenhouse because it removes surface contaminants.
2. Sometimes a brief alcohol rinse or swabbing with alcohol-wetted cheesecloth is appropriate especially with surfaces that are hairy or coated with thick wax.
3. Immerse the explants in the chlorine bleach solution, which should always be made up fresh. Always add 1-2 drops of Tween-20, detergent, or other wetting agent per 100 ml of bleach solution. A 10% bleach solution is prepared by adding 10 ml of chlorine bleach to a graduated cylinder and diluting with water to 100 ml. When disinfesting in culture tubes, pour the bleach solution into the culture tube cap and then into the culture tube; this helps to disinfest the cap. Place your hand over the tube and mix. Cap and agitate occasionally for 5 to 30 min for disinfestations. Commercial chlorine bleach once opened can age and lose its effectiveness. For this reason, small containers of commercial bleach are preferable to large ones.

4. Decant the bleach solution and rinse the explant in sterile water three to five times. This step is carried out in a transfer hood.

Surface Sterilization Protocol for Axillary Buds in Apple (Malus x domestica Borkh.)

Materials Required:

- Healthy, disease-free apple shoots with axillary buds
- Sterile distilled water
- Tween-20 (surfactant)
- 70% ethanol
- Sodium hypochlorite (NaOCl) solution (commercial bleach, 4 to 6% available chlorine)
- Mercuric chloride ($HgCl_2$) solution (0.1% w/v) – optional; handle with extreme care
- Sterile forceps and scalpels
- Laminar airflow hood
- Autoclaved filter paper
- Sterile Petri dishes

Procedure:

1. Collection and Pre-cleaning:

- Select vigorous, healthy apple shoots containing axillary buds.
- Trim shoots to ~2 to 3 cm segments containing the bud.
- Wash the explants under running tap water for 20 to 30 minutes to remove dust and surface debris.

2. Detergent Wash:

- Immerse explants in sterile distilled water containing 2 to 3 drops of Tween-20.
- Gently agitate for 10 to 15 minutes to loosen microbial films.
- Rinse thoroughly 3 to 5 times with sterile distilled water to remove detergent residues.

3. Surface Sterilization Steps:

These steps are to be performed inside the laminar airflow cabinet:

a) Ethanol Treatment:

- Dip explants quickly in 70% ethanol for 30 to 60 seconds. Immediately rinse once with sterile distilled water.

b) Sodium Hypochlorite Treatment:

- Transfer explants into a freshly prepared 0.5 to 1.0% NaOCl solution (adjusted from commercial bleach) containing 1 to 2 drops of Tween-20.
- Soak explants for 10 to 15 minutes with gentle intermittent shaking. Stronger NaOCl (up to 2%) can be used for highly contaminated field samples but shorter exposure (~5 to 7 minutes) is recommended to avoid tissue damage.

c) Optional - Mercuric Chloride Treatment (for stubborn contamination):

- Transfer explants into 0.1% $HgCl_2$ solution for 2 to 5 minutes with gentle agitation.
- Followed by at least 5 thorough rinses with sterile distilled water to completely remove any traces of $HgCl_2$. Caution: Mercuric chloride is highly toxic. Proper disposal protocols must be strictly followed.

4. Final Rinse and Drying:

- Perform 3 to 5 final rinses with sterile distilled water to ensure removal of residual sterilants.
- Blot dry explants gently on sterile autoclaved filter paper inside a sterile Petri dish.

CHAPTER 10

CALLUS INDUCTION AND PLANT REGENERATION FROM LEAF EXPLANTS OF APPLE (*MALUS X DOMESTICA* BORKH.)

Aim

To establish an efficient protocol for inducing callus and regenerating plants from leaf explants of apple under controlled in vitro conditions.

Principle

Callus is defined as a mass of "unorganized" plant cells. Callus can occur as a result of a plant wound. In plant tissue culture, it usually arises from pieces of tissue which the tissue culturist places on a medium containing certain plant growth regulators. When the tissue culturist moves the callus onto a new medium with different plant growth regulators, the cells in the callus are induced to organize and form new plant organs.

Callus formation is central to many investigative and applied tissue culture procedures. Callus can be multiplied and later used to clone numerous whole plants. Additionally, various genetic engineering protocols employ callus initiation procedures followed by genetic transformation. Then callus is regenerated in whole plant leads to formation of transgenic plants. Callus is also being generated for use in biotechnological procedures such as the formation of suspension cultures from which valuable plant products can be harvested.

Callus induction: Explants from several parts of plants can be used for callusing. The most successful explants are often young tissues of one or a few cell types. Pith cells of young stem are usually a good source of explant material. Initially, callus cells proliferate without differentiating, but eventually differentiation occurs within the tissue mass. Actively dividing cells are those uppermost and peripheral in the callus. The extent of overall differentiation usually depends on the hormone balance of the support medium and the physiological state of the tissue.

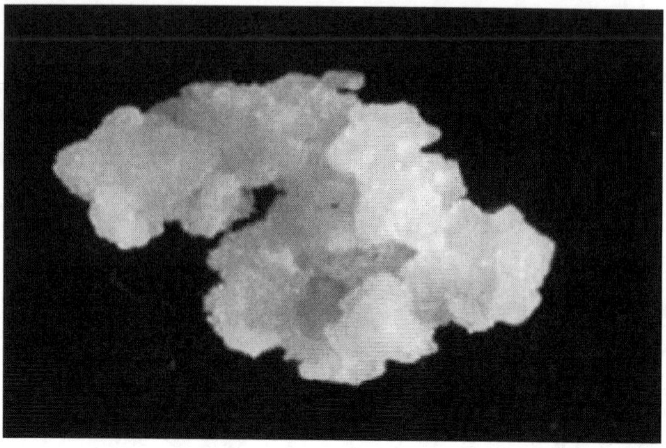

Callus tissue generated from leaf explants

Callus Multiplication: Actively growing callus can be initiated on culture media with an even physiological balance of cytokinin and auxin. After callus biomass increases two to four times (after 2 to 4 weeks of growth), callus can be divided and placed on fresh Callus Initiation Medium for callus multiplication. Multiplication procedures can be repeated several times (up to eight sequential transfers) before gross chromosome instability (or contamination) occurs.

Differentiation and Plant Regeneration: Multiplied callus can be stimulated to form shoots by increasing the cytokinin concentration and decreasing auxin content of culture media. Shoot masses can be cut apart and transferred to rooting medium. Once rooted, regenerated plants can be acclimatized to natural rather than "in vitro" growth conditions. Regenerated plants are especially valuable if the parent plant was itself unique or if the plants were genetically engineered. If, for example, multiplied callus was first used to form suspension cultures on which genetic engineering or cell selection was accomplished, resultant regenerated plants via tissue culture could possess special traits or capabilities.

Materials Required

- Healthy apple explants i.e., leaves
- Murashige and Skoog (MS) basal medium
- Plant Growth Regulators (PGRs):
  - » Auxins: 2,4-Dichlorophenoxyacetic acid (2,4-D), Indole-3-acetic acid (IAA)
  - » Cytokinins: 6-Benzylaminopurine (BAP), Thidiazuron (TDZ)

- Agar (gelling agent)
- Sucrose (carbon source)
- Distilled water (sterile)
- Instruments: Laminar airflow cabinet, Autoclave, Growth chamber or culture room, Forceps, scalpels, blades (sterile), Sterile Petri dishes, Culture tubes/jars
- Sterilization Agents: 70% ethanol, Sodium hypochlorite solution (NaOCl)

Protocol in Detail

1. Plant Materials and Explants Preparation

Collect young, healthy, fully expanded apple leaves from mother plants. Wash thoroughly under running tap water for 20-30 minutes. Then in a laminar airflow hood dip leaves in 70% ethanol for 30 seconds.

Transfer to 0.5-1% NaOCl solution with 2 to 3 drops of Tween-20 for 10-12 minutes with gentle shaking. Rinse 4-5 times with sterile distilled water. Blot dry on sterile filter paper before use.

2. Preparation of Culture Media:

Callus Induction Medium

- Weigh MS basal medium powder according to manufacturer's instructions and dissolve it in approximately 80% of the final volume of distilled water.
- Add 30 g/L (3%) sucrose and 7 g/L (0.7%) agar to the solution.
- Supplement the medium with 2,4-Dichlorophenoxyacetic acid (2,4-D) at a concentration of 1.0 to 2.0 mg/L for callus induction. If required, add 6-Benzylaminopurine (BAP) at 0.5 to 1.0 mg/L to improve callus quality.
- Stir the solution thoroughly until all components are completely dissolved and adjust the pH of the medium to 5.7 to 5.8 using 1N NaOH or 1N HCl as needed.
- Make up the volume to the desired final volume with distilled water. Dispense the medium into suitable containers. Autoclave at 121°C under 15 psi pressure for 15 to 20 minutes.
- If any heat-sensitive plant growth regulators are required, sterilize them separately using a 0.22 µm membrane filter and add them to the medium after it has cooled to approximately 45 to 50°C.

Shoot Induction Medium:

- Prepare MS basal medium as described above. Add 30 g/L sucrose and 7 g/L agar. Supplement with BAP at 0.5 to 2.0 mg/L and Indole-3-acetic acid (IAA) at 0.1–0.5 mg/L.

- Adjust pH to 5.7 to 5.8, make up the volume, and autoclave. Add heat-sensitive components, if any, by filter sterilization after cooling. TDZ (0.5 to 1.0 mg/L) can be optionally tested along with BAP to enhance shoot organogenesis.

Rooting Medium:

- Prepare half-strength MS basal salts. Add 20–30 g/L (2 to 3%) sucrose and solidify with 7 g/L agar. Supplement with Indole-3-butyric acid (IBA) at 0.5 to 1.0 mg/L. Adjust pH to 5.7 to 5.8, autoclave at 121°C and 15 psi for 15 to 20 minutes, and cool before use.

3. Procedure:

A) Callus Induction and Multiplication

- Cut sterilized leaves into small pieces, avoiding midribs (only lamina sections are preferable).
- Place leaf segments abaxial side down on the callus induction medium.
- Incubate cultures in dark conditions at 25±2°C. Monitor for callus formation over 2-3 weeks.

- Once friable or compact creamy white/yellowish callus typically develops, Subculture the induced callus onto fresh callus induction medium every 3-4 weeks to promote growth and maintain vitality.
- Select healthy, rapidly growing callus masses for further regeneration experiments.

B) Shoot Regeneration

- Transfer healthy callus pieces onto shoot induction medium.
- Incubate cultures under 16-hour photoperiod (light intensity ~40-50 µmol m^{-} s^{-}) at 25±2°C.
- Shoots usually initiate within 3-6 weeks, then Subculture proliferating shoots every 2-3 weeks onto fresh regeneration medium.

C) Rooting of Shoots

- Excise regenerated shoots (~2-3 cm long) and transfer them to rooting medium (half-strength MS + IBA).
- Incubate under standard culture conditions. Roots usually form within 2-4 weeks.

D) Acclimatization & Hardening

- Gently remove plantlets from the culture medium and wash off agar residues with sterile water.
- Transfer rooted plantlets to sterile potting substrate (e.g., peat moss: vermiculite 1:1).
- Cover with transparent plastic covers or keep inside a mist chamber to maintain high humidity.
- Gradually harden over 2-4 weeks before transferring to the greenhouse.

Potential Problems Encountered During Callus Culture and Their Troubleshooting

- Lack of callus formation can result from physiologically inactive or mature explant tissues and improper hormone balance, which can be overcome by using juvenile leaf explants and optimizing 2,4-D (1 to 2 mg/L) with low BAP (0.5 to 1.0 mg/L), along with optional antioxidant pretreatment.
- Browning of explants due to phenolic oxidation can be minimized by pre-treating with ascorbic or citric acid (50 to100 mg/L), incorporating activated charcoal in the medium, and practicing frequent subculturing.
- Contamination arising from inadequate surface sterilization or aseptic handling can be controlled by extending NaOCl treatment (12 to 15 min with Tween-20), ensuring thorough rinsing, and maintaining strict laminar hood protocols.
- Soft, mucilaginous callus formation typically caused by high auxin levels or excessive humidity can be addressed by reducing 2,4-D, slightly increasing agar concentration, and improving ventilation within culture vessels.

- Poor shoot regeneration from callus often results from hormonal imbalance or non-competent callus and can be improved by increasing cytokinin levels (BAP or TDZ), reducing auxin, selecting friable, greenish callus, and ensuring a proper light regime.
- Failure of rooting in regenerated shoots may occur due to low auxin levels or immature shoot tissues and can be resolved by increasing IBA (1.0 to 1.5 mg/L), using half-strength MS salts, and maintaining a cool, stable environment.
- Wilting or death of plantlets post-transfer is commonly due to abrupt environmental transition and weak roots, which can be mitigated through gradual acclimatization under high humidity, use of sterile porous substrates, and provision of indirect light.

CHAPTER 11

DIRECT REGENERATION (DIRECT SHOOT ORGANOGENESIS) FROM EXPLANTS OF APPLE (*MALUS* X *DOMESTICA* BORKH.)

Aim

To achieve direct shoot regeneration from apple explants under in vitro conditions, bypassing callus formation, for efficient micropropagation and genetic stability.

Principle

Direct regeneration in plant tissue culture is based on the ability of differentiated somatic cells to revert to a meristematic state and directly form organs, such as shoots or roots, without passing through an intermediate callus phase. This process preserves the genetic stability of the regenerated plants by minimizing somaclonal variation, which is more common when regeneration occurs via callus. Direct regeneration is typically induced by carefully manipulating the balance of plant growth regulators, especially cytokinins and auxins, in the culture medium. A higher cytokinin-to-auxin ratio generally promotes shoot initiation, while an elevated auxin concentration favours root development. Environmental factors such as light intensity, temperature, and photoperiod also play crucial roles in guiding direct organogenesis. Through appropriate explant selection, hormonal regulation, and controlled culture conditions, direct regeneration enables efficient, rapid, and true-to-type propagation of plants.

Direct Shoot Organogenesis

Materials Required

- Healthy apple explants: apical buds, Axillary buds, shoot tips, nodal segments, cotyledons
- Murashige and Skoog (MS) basal medium
- Plant Growth Regulators (PGRs):
    - » Cytokinins: 6-Benzylaminopurine (BAP), Thidiazuron (TDZ)
    - » Auxins: Indole-3-acetic acid (IAA) or Naphthaleneacetic acid (NAA)
- Sucrose (3%)
- Agar (0.7%)
- Distilled water (sterile)
- 1N NaOH and 1N HCl (for pH adjustment)
- Sterilization Agents: 70% ethanol, Sodium hypochlorite solution (NaOCl)
- Instruments: Laminar airflow cabinet, Culture tubes, jars, Petri dishes, Autoclave, Growth chamber, Forceps, scalpels, blades (sterile), pH meter

Protocol in Detail

A) Surface Sterilization of Explants:

- Wash collected explants under running tap water for 20-30 minutes. Transfer explants into sterile distilled water containing 2-3 drops of Tween-20 and shake for 10 minutes. Rinse explants thoroughly with sterile distilled water.
- Dip in 70% ethanol for 30 seconds. Then surface sterilize with 0.5-1.0% sodium hypochlorite (NaOCl) solution for 10-15 minutes with gentle shaking.

- Rinse 3-5 times with sterile distilled water. Blot dry using sterile filter paper inside a laminar airflow cabinet.

B) Preparation of Media:

Shoot Induction Medium:

- Weigh the required amount of MS basal salts and dissolve in approximately 80% of the final volume of distilled water.
- Add 30 g/L (3%) sucrose and 7 g/L (0.7%) agar to the solution. Stir until completely dissolved.
- Now add 6-Benzylaminopurine (BAP) at a concentration of 1.0-2.0 mg/L or Thidiazuron (TDZ) at 0.2-0.5 mg/L for shoot induction.
- If shoot elongation needs to be enhanced, optionally add Indole-3-acetic acid (IAA) or Naphthaleneacetic acid (NAA) at 0.1-0.5 mg/L.
- Adjust the pH of the medium to 5.7-5.8 using 1N NaOH or 1N HCl.
- Make up the final volume with distilled water.
- Dispense into culture vessels and autoclave at 121°C for 15-20 minutes at 15 psi pressure.

Note: If any plant growth regulators are heat-sensitive, filter-sterilize them separately using a 0.22 µm filter and add them to the cooled medium (around 45-50°C).

Rooting Medium:

- Prepare half-strength MS basal salts solution.
- Add 20-30 g/L (2-3%) sucrose and 7 g/L agar.
- Supplement the medium with Indole-3-butyric acid (IBA) at a concentration of 0.5-1.0 mg/L for root induction.
- Stir until fully dissolved, adjust the pH to 5.7-5.8, and make up the final volume.
- Autoclave at 121°C and 15 psi for 15-20 minutes. Cool the medium before use.

C) Procedure:

- Under sterile conditions, excise explants into appropriate sizes depending on the type of explant (e.g., nodal segments, leaf discs).
- Place the explants either horizontally or vertically onto the shoot induction medium, depending on their orientation requirement.
- Incubate cultures at 25±2°C under a 16-hour photoperiod with a light intensity of 40–50 µmol m⁻ s⁻ .

- Monitor explants over 2 to 4 weeks for signs of direct shoot initiation. Successful direct regeneration is indicated by the direct emergence of shoot buds from the explant without any intervening callus formation.
- Transfer developing shoots to fresh MS medium supplemented with a lower cytokinin concentration (e.g., BAP 0.2-0.5 mg/L) to promote shoot elongation.
- Maintain the elongated shoots under the same controlled environmental conditions. Excise elongated shoots (~2-3 cm in length) and transfer them to rooting medium (half-strength MS supplemented with IBA).
- Expect root initiation within 2-4 weeks under standard light and temperature conditions.
- Carefully remove rooted plantlets from the culture medium and wash off residual agar with sterile distilled water.
- Transfer plantlets to pots containing a sterilized potting mixture (peat moss: sand: vermiculite in a 1:1:1 ratio).
- Cover the pots with transparent plastic bags or place them inside a humidity chamber to maintain high humidity.
- Gradually acclimate the plantlets to greenhouse conditions over a period of 2-4 weeks by progressively reducing humidity.

Potential Problems Encountered During Direct Regeneration and Their Troubleshooting

- Low shoot induction frequency due to unresponsive explants can be improved by selecting actively dividing tissues such as shoot tips or nodal segments and optimizing cytokinin levels, particularly BAP (1.0 to 2.0 mg/L) or TDZ (0.2 to 0.5 mg/L).
- Phenolic leaching and browning of explants, which hinder regeneration, can be mitigated by pretreatment with antioxidants like ascorbic acid and frequent subculturing onto fresh media.
- Contamination arising from insufficient surface sterilization can be controlled by extending exposure to sodium hypochlorite (0.5 to 1%) with surfactant, followed by multiple rinses in sterile distilled water.
- Shoot bud formation with intervening callus, instead of direct organogenesis, indicates hormonal imbalance and can be corrected by increasing cytokinin-to-auxin ratio to suppress callus and promote direct shoot initiation.
- Short or stunted shoots can result from low elongation efficiency and may be overcome by transferring regenerating shoots to medium with reduced cytokinin (e.g., BAP 0.2 to 0.5 mg/L) and optional low auxin (IAA or NAA).

- Rooting failure in elongated shoots is often due to suboptimal auxin concentration and can be addressed by using half-strength MS medium supplemented with IBA (0.5 to 1.0 mg/L).
- Wilting or mortality during acclimatization occurs due to sudden humidity drop and weak root systems, which can be avoided by gradual hardening in high-humidity environments using transparent covers and sterile potting mix.

CHAPTER 12

SOMATIC EMBRYOGENESIS IN APPLE (*MALUS* X *DOMESTICA* BORKH.)

Aim

To induce and regenerate somatic embryos from cotyledons of immature zygotic embryos of apple for efficient plantlet development.

Principle

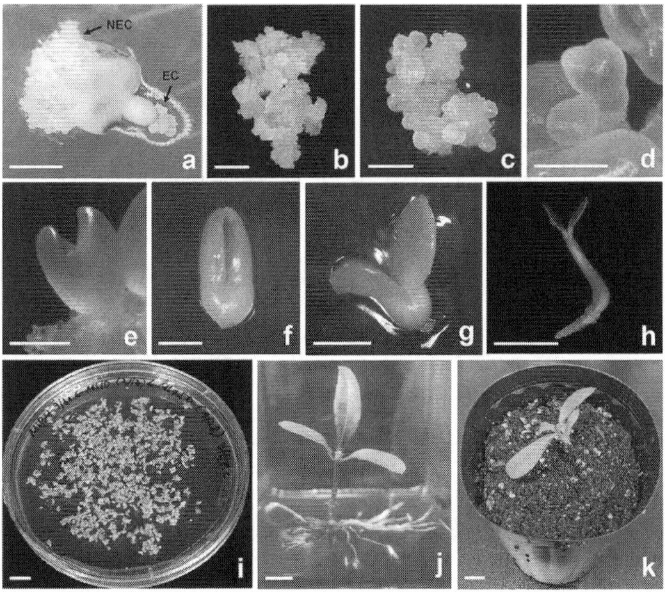

Somatic embryogenesis is the process by which somatic cells are induced to form embryos that can develop into complete plants. In apple, this is typically achieved by culturing explants like immature cotyledons on auxin-rich media (e.g., with 2,4-D or NAA) under dark conditions. The somatic embryos formed are later transferred to a development medium with reduced or no hormones to promote maturation and plant regeneration. This technique is valuable for clonal propagation, genetic transformation, and virus-free plant production.

Materials Required

- Cotyledons from immature zygotic embryos (collected from <60-day-old fruits)
- Murashige and Skoog (MS) basal medium
- Plant Growth Regulators (PGRs): Naphthaleneacetic acid (NAA), 6-Benzylaminopurine (BAP), Kinetin (KIN), Indole-3-butyric acid (IBA)
- Organic supplements: Adenine sulfate, L-glutamine, Glycine, Casein hydrolysate
- Sucrose
- Agar
- Distilled water
- Sterilization agents (70% ethanol, 0.5-1.0% NaOCl)
- Instruments: Laminar airflow cabinet, Growth chamber, Autoclave, Sterile Petri dishes, culture tubes, flasks

Protocol in Detail

A) Preparation of Media:

For Induction Medium (IM):

- Dissolve MS basal salts in 80% of the final volume of distilled water.
- Add 60 g/L sucrose and 7 g/L agar; stir until completely dissolved.
- Add 6.0 mg/L NAA, 500 mg/L casein hydrolysate, adenine sulfate (as per MS vitamins), 100 to 200 mg/L L-glutamine, and 2 mg/L glycine.
- Mix thoroughly, adjust pH to 5.7 to 5.8, make up to final volume, and autoclave at 121°C for 15–20 minutes.

For Development Medium (DM):

- Prepare MS basal salts, add 30 g/L sucrose and 7 g/L agar, supplement with 0.5 mg/L BAP and 0.05 mg/L NAA, adjust pH, and autoclave.

For Multiplication Medium (MM):

- Prepare MS basal salts with 30 g/L sucrose and 7 g/L agar, add 0.1 mg/L each of BAP, KIN, and IBA, adjust pH, and autoclave.

For Germination Medium:

- Use half-strength MS basal salts with 30 g/L sucrose and 7 g/L agar, without any plant growth regulators; adjust pH and autoclave.
- Cool all media to 45 to 50°C before pouring into sterile Petri dishes or culture vessels.

B) Surface Sterilization of Explants

- Immature apple fruits (approximately 60 days post-pollination) are collected, and seeds are carefully excised to isolate the cotyledons. The cotyledons are then washed thoroughly under running tap water for 20 to 30 minutes to remove surface contaminants.
- Inside a laminar airflow cabinet, the cotyledons are immersed in 70% ethanol for 30 seconds for preliminary sterilization.
- Following this, they are treated with 0.5 to 1.0% sodium hypochlorite (NaOCl) solution for 10 to 15 minutes to ensure surface sterilization.
- After the chemical treatment, the cotyledons are rinsed 3 to 5 times with sterile distilled water to remove all traces of sterilant.
- Finally, the sterilized cotyledons are blotted dry on sterile filter paper before inoculation onto culture media.

C) Procedure:

- Place sterilized cotyledons onto Induction Medium (IM) with the adaxial side in contact with the medium.
- Incubate cultures in complete darkness at 25±2°C for 20 days.
- Swelling and early formation of embryogenic structures will appear, after 20 days, transfer the explants onto Development Medium (DM).
- Incubate cultures under 16/8-hour light/dark photoperiod at 25±2°C.
- Monitor for formation of somatic embryos (globular, heart, torpedo stages).
- Mature embryos are transferred to hormone-free half-strength MS medium for cold treatment (4°C) for 2 months to promote conversion.
- After cold treatment, transfer embryos to fresh half-strength MS medium under normal growth conditions to promote shoot and root development.
- For repetitive somatic embryogenesis, subculture developing embryos onto Multiplication Medium (MM) containing low concentrations of BAP, KIN, and IBA. Maintain cultures in continuous darkness.

Acclimatization:

- Gently remove regenerated plantlets from culture vessels.
- Wash off agar residues with sterile distilled water.

- Transfer plantlets to pots containing sterilized potting substrate (peat moss: sand: vermiculite in 1:1:1 ratio).
- Cover with transparent plastic bags or place in a high-humidity chamber.
- Gradually expose plantlets to ambient greenhouse conditions over 2-4 weeks.

Potential Problems Encountered During Somatic Embryogenesis and Their Troubleshooting

- Low frequency of somatic embryo induction may result from physiological immaturity or poor responsiveness of cotyledons, which can be resolved by using explants from fruits harvested 50 to 60 days post-pollination and culturing them on auxin-rich induction medium (e.g., 6.0 mg/L NAA).
- Contamination of cotyledon explants during culture initiation is commonly due to inadequate surface sterilization, which can be prevented by optimizing exposure to 70% ethanol and 0.5 to 1.0% NaOCl, followed by multiple rinses in sterile water.
- Browning or necrosis of explants due to phenolic oxidation can be minimized by incorporating antioxidants (ascorbic acid or activated charcoal) and ensuring quick transfer to fresh media.
- Failure of embryogenic structure development under induction conditions often stems from inadequate dark incubation or insufficient organic supplements, which can be corrected by ensuring continuous darkness and enriching media with glutamine, casein hydrolysate, and adenine sulfate.
- Somatic embryos failing to mature or transition to advanced stages like torpedo or cotyledonary forms may be due to imbalanced PGRs or excess osmotic stress, and can be addressed by shifting to development medium with reduced hormones and appropriate sucrose concentration (30 g/L).
- Poor germination and conversion of embryos into plantlets may occur if cold treatment is omitted, thus enforcing a two-month chilling period at 4 C can significantly improve conversion efficiency.
- Rooting failure in somatic plantlets may be caused by residual hormonal effects or poor medium composition, which can be rectified by transferring embryos to half-strength MS medium without PGRs or with low IBA (0.5 to 1.0 mg/L).
- Acclimatization losses post-regeneration is typically due to sudden humidity shifts or microbial contamination in substrate, which can be mitigated through gradual hardening in high humidity using sterile peat-based substrates and protective covers.

CHAPTER 13

DOUBLED HAPLOID PRODUCTION IN APPLE (*MALUS × DOMESTICA* BORKH.) VIA ANTHER CULTURE AND POLLEN CULTURE

Aim

To produce doubled haploid (DH) apple plants via androgenesis using both anther and pollen culture routes, with optimized protocols for embryo induction, cold treatment, shoot regeneration, and rooting.

Principle

Doubled haploid production enables the generation of completely homozygous lines in a single generation, significantly accelerating breeding programs. In apple, both anther and pollen culture is employed to induce androgenesis, where immature microspores within cultured anthers undergo embryogenic development. Cold treatment and medium composition critically influence embryo induction and regeneration. Rooted and nonrooted embryos respond differently to shoot regeneration, and tailored hormone regimes are required for each. Chromosome doubling of haploid regenerants is then performed to produce fertile, true-breeding lines. Colchicine is applied to haploids to double their chromosome number, yielding stable DH lines

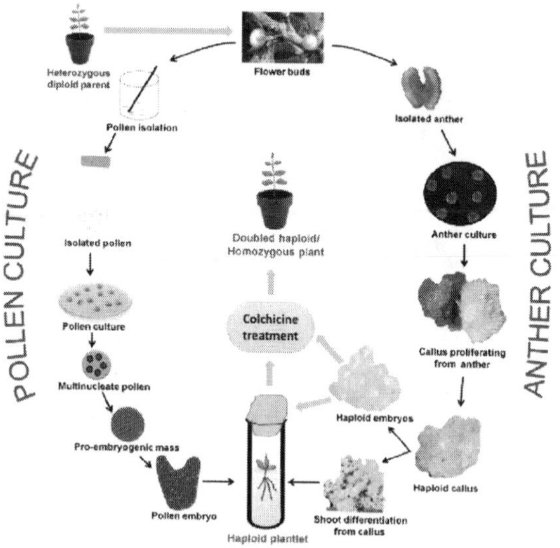

Materials Required

- Apple flower buds at "First Pink" stage
- N6 and MS basal salts
- Sucrose, Agar
- Colchicine (0.1% to 0.5% aqueous solution)
- Plant Growth Regulators: BAP, NAA, IBA, GA_3
- Sterilization agents: 70 % Ethanol, Sodium hypochlorite, sterile distilled water
- Instruments: Laminar airflow, scalpel, forceps, culture vessels, autoclave

Protocol in Detail

A) Preparation of Media:

Embryo Induction Medium Composition:

- N6 basal salts
- 10 µM BAP
- 0.5 µM NAA
- 5% sucrose
- 0.8% agar

Cold Treatment Medium Composition:

- ½ MS salts
- 1.5% sucrose
- 0.8% agar

Shoot Medium Composition:

For Rooted Embryos:

- ½ MS hormone-free medium with 1.5% sucrose, or
- ½ MS + 1 mg/L BAP + 5 µM GA

For Nonrooted Embryos:

- Use either MS or ½ MS medium with:
- 0.5-1.0 mg/L BAP
- 2 mg/L GA or 0.1 mg/L IBA (optional)
- 3% sucrose
- 0.65-0.8% agar

Shoot Multiplication Medium:

- MS basal salts
- 1 mg/L BAP
- 0.1 mg/L IBA
- 3% sucrose
- 0.65% agar

B) Procedure:

- Collect flower buds from selected apple cultivars at the "First Pink" stage or ensure microspores are at the appropriate developmental stage (e.g., uninucleate microspores for androgenesis) for optimal androgenic response.
- Place the collected buds in sterile Petri dishes and store them in complete darkness at 4 C for 25 days to enhance their embryogenic competence.
- Immerse flower buds in 70% ethanol for 30 seconds followed by sodium hypochlorite solution (1-2%) for 7-8 minutes.
- Rinse the buds three times with sterile distilled water and gently blot them dry with sterile filter paper.

For Anther culture:

- Using sterile tools, dissect the buds and carefully isolate the anthers. Place the anthers onto embryo induction medium with the pH adjusted to 5.8 before autoclaving.
- Incubate the cultures in complete darkness at 25 C and monitor for embryo formation over a period of 7 to 18 weeks.
- Select well-formed embryos measuring 3 to 8 mm in size and transfer them onto cold treatment medium without hormones with the pH adjusted to 5.8 before autoclaving.

- Keep the embryos in darkness at 4°C for 12 to 20 weeks to enhance their shoot regeneration capacity.
- Sort embryos into rooted and nonrooted categories before shoot regeneration to apply the appropriate culture media.
- After cold treatment, transfer rooted embryos to hormone-free ½ MS medium containing 1.5% sucrose. For rooted embryos use Shoot Medium for Rooted Embryos and for nonrooted embryos given above with the pH adjusted to 5.8 before autoclaving.
- Incubate all cultures at 25°C with a 16/8 h light/dark photoperiod and subculture every 4 weeks.
- Once shoots develop, transfer them to shoot multiplication medium.
- Maintain cultures under a 16-hour light period at 25°C and subculture regularly to enhance shoot proliferation.
- Excise elongated shoots (>10 mm) and place them on rooting medium.
- Incubate the cultures in the dark at 25°C for 7 to 20 days, then shift them to light conditions (16/8 h photoperiod) for continued growth.

For Pollen Culture:

- Isolate pollen grains by crushing anthers in sterile water
- Centrifuge and culture pollen on induction medium. Observe multinucleate pollen → pro-embryogenic masses → pollen embryos
- Transfer to test tubes with rooting medium
- Generate haploid plantlets, followed by colchicine treatment to induce DH
- Regenerate shoots and roots, then acclimatize

Colchicine Treatment (Common Step)

- Treat haploid plantlets with 0.1–0.5% colchicine for 24–48 hours (dip root zone or culture on colchicine-supplemented medium)
- Rinse thoroughly and transfer to MS medium for recovery

C) Acclimatization:

- Gently wash rooted plantlets with sterile distilled water to remove agar and transfer them to pots filled with sterilized vermiculite (60 60 mm).
- Cover the pots with transparent polyethylene bags to maintain high humidity and gradually remove the covers over 2 to 3 weeks to allow adaptation to ambient conditions.

D) Ploidy Confirmation:

- After acclimatization, confirm the ploidy status of regenerated plants using flow cytometry to assess nuclear DNA content.

- Additionally, marker analysis can be performed with primers specific to the donor cultivar to verify their haploid origin.

Potential Problems Encountered During Doubled Haploid Production and Their Troubleshooting

- Low embryo induction efficiency from anthers or pollen is often due to inappropriate microspore stage or insufficient cold pre-treatment, which can be resolved by selecting buds at the 'First Pink' stage and applying a 25-day cold incubation at 4°C.
- Microbial contamination during culture initiation arises from incomplete sterilization, and can be prevented by using optimized disinfection protocols involving 70% ethanol and 1 to 2% NaOCl followed by sterile water rinses.
- Failure to form embryos in anther cultures may be due to hormonal imbalance or media composition, which can be addressed by optimizing the N6-based embryo induction medium with BAP (10 µM) and NAA (0.5 µM).
- Poor embryo maturation or degeneration during culture is often the result of osmotic stress or prolonged darkness, which can be corrected by regulating sucrose concentrations and transferring to cold treatment media at the right time.
- Uneven regeneration from rooted vs. non-rooted embryos stems from applying a uniform protocol, which should be corrected by using distinct shoot media tailored for each group, such as BAP + GA for rooted and BAP + IBA for non-rooted embryos.
- Callus formation instead of embryo development in pollen cultures is typically due to damaged pollen or incorrect induction media, which can be improved by gentle anther crushing, proper pollen isolation, and quick inoculation onto BAP-rich medium.
- Colchicine toxicity or failure to induce chromosome doubling results from incorrect concentration or exposure duration, which can be resolved by limiting colchicine to 0.1–0.5% and applying it for 24 to 48 hours only to root tips or culture medium.
- Rooting failure in regenerated shoots often occurs due to hormonal carryover or media incompatibility, which can be managed by shifting to hormone-free or IBA-supplemented half-strength MS media.
- High mortality during acclimatization is linked to sudden humidity loss or microbial contamination in substrates, and can be mitigated by gradual humidity reduction using polyethylene bags and sterile, well-draining potting mix.

- Inaccurate identification of haploids or DH lines due to somatic contamination can be avoided by confirming ploidy through flow cytometry and validating origin with cultivar-specific molecular markers.

CHAPTER 14

PROTOPLAST FUSION IN APPLE (*MALUS* × *DOMESTICA* BORKH.)

Aim

To develop and implement a protocol for protoplast fusion in apple, facilitating somatic hybridization and enabling the combination of desirable traits from different cultivars or species.

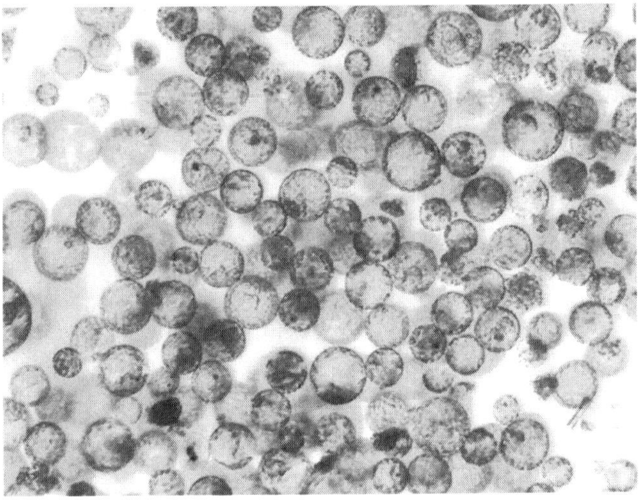

Principle

Protoplast fusion involves the merging of two distinct protoplasts, resulting in a hybrid cell that combines genetic material from both parent cells. This technique allows for the creation of novel genetic combinations, especially between sexually incompatible species or cultivars. The fused protoplasts can regenerate into whole plants, offering a pathway for introducing new traits and enhancing genetic diversity.

Materials Required

- Young, healthy leaves or embryogenic callus from selected apple cultivars.
- Enzymes for Protoplast Isolation: Cellulase Onozuka R-10, Macerozyme R-10
- Osmoticum: Mannitol
- Fusion Agents: Polyethylene glycol (PEG) 6000, Calcium chloride ($CaCl_2$)
- Culture Media: Modified Murashige and Skoog (MS) medium
- Plant growth regulators: 2,4-Dichlorophenoxyacetic acid (2,4-D), Benzylaminopurine (BAP), Naphthaleneacetic acid (NAA)
- Equipment: Laminar airflow hood, Sterile Petri dishes, Centrifuge, Hemocytometer, Incubator

Protocol in Detail

A) Preparation of Media:

1. Enzyme Solution for Protoplast Isolation

- Cellulase Onozuka R-10: 1.5%
- Macerozyme R-10: 0.5%
- Mannitol: 0.4 M
- $CaCl_2$: 10 mM
- BSA: 0.1%

2. Washing Solution

- Mannitol: 0.4 M
- $CaCl_2$: 10 mM

3. Fusion Solution

- PEG 6000: 40% (w/v)
- $CaCl_2$: 0.1 M
- Mannitol: 0.4 M

4. Protoplast Culture Medium

- Basal medium: Modified MS
- 2,4-D: 0.5 mg/L
- BAP: 0.5 mg/L
- Mannitol: 0.4 M
- Agar (if solidified): 0.7–0.8%

5. Regeneration Medium

- Basal medium: MS
- BAP: 0.5 mg/L
- NAA: 0.1 mg/L
- Mannitol: 0.2 M

6. Rooting Medium (optional)

- ½ MS
- IBA: 0.5 mg/L
- Sucrose: 3%
- Agar: 0.65%

B) Procedure:

- Select young, fully expanded leaves or embryogenic callus from in vitro-grown apple plants.
- Rinse the tissue with sterile distilled water.
- Surface sterilize using 70% ethanol for 30 seconds, followed by 0.1% sodium hypochlorite solution for 10 minutes. Then rinse three times with sterile distilled water.
- Cut tissue into 1 mm pieces, immerse in enzyme solution with pH adjusted to 5.8 (adjust before filter sterilization), and incubate in the dark at 25°C for 12 to 16 hours with gentle shaking.
- Filter digested mixture through a 40 µm mesh and collect protoplasts by centrifugation (100 g, 5 min).
- Resuspend the protoplasts in washing solution with pH adjusted to 5.8 and repeat the centrifugation step twice.
- Adjust the protoplast density to 1 10^6 protoplasts/mL using the washing solution.
- Mix equal volumes of two protoplast suspensions for fusion. Add fusion solution with pH adjusted to 5.8 dropwise and incubate at room temperature for 20 min.
- Gradually dilute the fusion mixture with washing solution over 30 minutes to remove PEG.
- Centrifuge at 100 g for 5 minutes and resuspend the fused protoplasts in culture medium.
- Embed the fused protoplasts in a thin layer of alginate on Petri dishes or culture them in protoplast culture liquid medium with pH adjusted to 5.8

- Incubate in the dark at 25°C for the first 7 days to promote cell wall regeneration and initial divisions.
- After 7 days, transfer cultures to a light regime (16-hour photoperiod) to encourage callus formation.
- Once calli are visible (typically after 3-4 weeks), transfer them to regeneration medium.
- Incubate under a 16-hour photoperiod at 25°C.
- Monitor for shoot and root development over the next 4-6 weeks.
- Once plantlets are sufficiently developed, transfer them to rooting medium and subsequently acclimatize them in soil.

Potential Problems Encountered During Protoplast Fusion and Their Troubleshooting

- Low yield of viable protoplasts is often due to inefficient enzymatic digestion, which can be improved by optimizing enzyme concentrations (Cellulase 1.5%, Macerozyme 0.5%), incubation time (12 to 16 hrs), and maintaining gentle agitation in the dark.
- Excessive protoplast lysis during isolation is usually caused by osmotic imbalance or mechanical damage, and can be minimized by using 0.4 M mannitol as an osmotic stabilizer and avoiding harsh pipetting or centrifugation steps.
- Clumping of protoplasts during PEG treatment may result in poor fusion efficiency, which can be resolved by gently mixing PEG solution dropwise and gradually diluting it with washing solution to minimize toxicity and aggregation.
- Failure of fusion between protoplasts can occur due to suboptimal PEG concentration or calcium levels, which can be corrected by using 40% PEG 6000 with 0.1 M $CaCl_2$ in 0.4 M mannitol and maintaining room temperature during fusion.
- Poor cell wall regeneration post-fusion may result from inadequate osmotic support or suboptimal culture conditions, which can be overcome by maintaining mannitol concentration at 0.4 M and incubating in the dark for the initial 7 days.
- Limited cell division and callus formation are often due to imbalanced hormone composition or poor protoplast viability, and can be improved by using modified MS medium supplemented with balanced concentrations of 2,4-D and BAP.

- Somatic hybrid calli failing to regenerate shoots may result from prolonged exposure to high auxin levels or lack of cytokinin stimulus, and can be addressed by transferring calli to MS regeneration medium containing BAP (0.5 mg/L) and NAA (0.1 mg/L).
- Rooting failure of regenerated shoots may occur due to hormonal carryover or weak physiological state, which can be corrected by culturing on half-strength MS medium supplemented with IBA (0.5 mg/L) and maintaining optimal humidity.
- Acclimatization losses are typically due to abrupt environmental changes or microbial contamination, which can be minimized by gradual humidity reduction, use of sterile potting mix, and protective transparent covers during early stages.

CHAPTER 15

EMBRYO RESCUE IN SEEDLESS GRAPE (*VITIS VINIFERA*)

Aim

To rescue and regenerate viable seedless grape plantlets from immature embryos, particularly those resulting from interspecific crosses, by employing in vitro culture techniques to overcome post-zygotic barriers.

Principle

Embryo rescue is an in vitro technique used to cultivate immature or weak embryos that might otherwise abort due to incompatibility barriers in conventional breeding. By excising embryos at specific developmental stages and culturing them on nutrient media supplemented with appropriate growth regulators, it is possible to promote their development into viable plantlets.

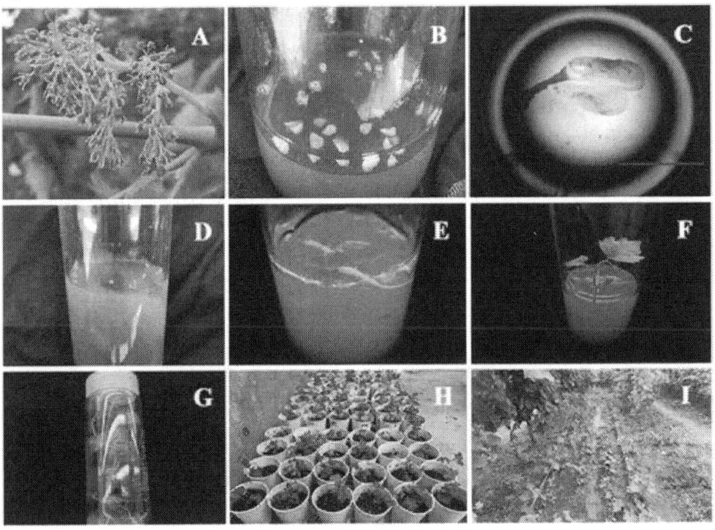

Materials Required

- Immature fruits of grape cultivars harvested 40 to 60 days after pollination.
- Murashige and Skoog (MS) basal salts and vitamins
- Sucrose, Maltose, Casein hydrolysate, Myo-inositol, Agar
- Plant Growth Regulators (PGRs): Indole-3-acetic acid (IAA), Gibberellic acid (GA_3), Kinetin
- Sterilization Agents: 70% Ethanol, Sodium hypochlorite solution (1.25%)
- Equipment: Laminar airflow cabinet, Scalpel and forceps, Culture vessels (e.g., Petri dishes, test tubes), Autoclave, Growth chamber

Protocol in Detail

A) Preparation of Media:

Embryo Culture Medium:

- MS basal salts and vitamins
- Sucrose: 3% (w/v)
- Maltose: 3% (w/v)
- Casein hydrolysate: 500 mg/L
- Myo-inositol: 100 mg/L
- Agar: 6 g/L

Embryo Germination Medium:

- MS basal salts and vitamins
- Sucrose: 2% (w/v)
- Casein hydrolysate: 500 mg/L
- Agar: 6 g/L

Plantlet Development Medium:

- MS basal salts and vitamins
- Sucrose: 2% (w/v)
- Agar: 6 g/L

B) Procedure:

- Harvest immature fruits 40 to 60 days after pollination, depending on the developmental stage of the embryo.
- Immerse fruits in 70% ethanol for 30 seconds. Transfer to 1.25% sodium hypochlorite solution for 15 minutes. Rinse three times with sterile distilled water.

- Under sterile conditions, dissect seeds from sterilized fruits. Carefully excise embryos using a scalpel and forceps, minimizing damage.
- Place excised embryos onto the prepared embryo culture medium with pH adjusted to 5.8 before autoclaving.
- Incubate cultures at 25 C with a 16-hour photoperiod.
- Monitor embryos for signs of germination over 2 to 4 weeks. Once germination is evident, transfer embryos to the embryo germination medium with pH adjusted to 5.8 before autoclaving.
- Continue incubation under the same environmental conditions.
- Upon development of shoots and roots, transfer plantlets to the plantlet development medium with pH adjusted to 5.8 before autoclaving (Adjustments to PGR concentrations may be necessary based on specific cultivar responses).
- Maintain cultures until plantlets reach a suitable size for acclimatization.

C) Acclimatization:

- Gently remove plantlets from the medium and wash off agar residues.
- Transfer plantlets to pots containing sterilized soil or vermiculite.
- Cover with transparent plastic to maintain high humidity.
- Gradually acclimate to ambient conditions over 2 to 3 weeks.

Potential Problems Encountered During Embryo Rescue and Their Troubleshooting

- Failure of embryo excision or mechanical damage during isolation is often due to immature handling techniques or improper tools, which can be prevented by using fine, sterile scalpels under magnification and working swiftly to avoid desiccation.
- Embryo browning and necrosis post-excision may result from phenolic oxidation, and can be minimized by using antioxidant pre-treatments like ascorbic acid or by transferring embryos quickly to the culture medium.
- Low germination rates are frequently due to harvesting embryos too early or poor medium composition, which can be improved by precisely timing embryo collection (40–60 days post-pollination) and using enriched media containing maltose, casein hydrolysate, and myo-inositol.
- Callus formation instead of shoot development may occur due to excess auxin or hormonal imbalance, and can be corrected by reducing IAA concentration and supplementing with kinetin or GA_3 to encourage direct germination.

- Abnormal shoot or root development often results from suboptimal photoperiod or PGR regime, which should be corrected by maintaining a 16-hour photoperiod and optimizing PGRs specific to the cultivar's regeneration response.
- Plantlets failing to acclimatize after transfer may be due to sudden humidity shock or poor root system, which can be resolved by gradual humidity reduction and using well-draining, sterile substrates like vermiculite or a peat mix.
- Contamination of embryo cultures can occur due to incomplete surface sterilization of fruits, which is best prevented by combining ethanol and NaOCl treatments followed by multiple rinses with sterile distilled water.

CHAPTER 16

IN VITRO MERISTEM CULTURE-A STRATEGY FOR VIRUS-FREE APPLE (*MALUS × DOMESTICA* BORKH.) PLANT PRODUCTION

Aim

To establish a reliable in vitro meristem culture protocol for the production of virus-free apple plants, particularly targeting vegetatively propagated cultivars susceptible to viral infections.

Principle

Meristem culture exploits the fact that apical meristems are often free or have lower concentrations of viruses due to limited vascular connections. By excising and culturing these meristems under sterile conditions, it is possible to regenerate whole plants devoid of systemic viral infections. Combining meristem culture with thermotherapy enhances the elimination of viruses by inhibiting their replication and movement.

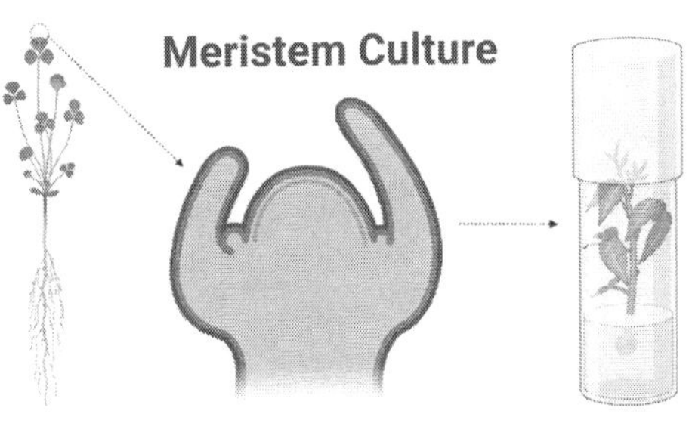

Materials Required

- Apple shoots from infected 'Oregon Spur-II' trees.
- Sterilization Agents:
- 70% Ethanol.
- 2% Sodium hypochlorite solution.
- Murashige and Skoog (MS) basal salts and vitamins.
- Sucrose.
- Agar.
- Plant Growth Regulators (PGRs): Benzyl adenine (BA), Gibberellic acid (GA_3), Indole-3-butyric acid (IBA) or Naphthaleneacetic acid (NAA) or Indole-3-acetic acid (IAA).
- Polyvinylpyrrolidone (PVP) or Ascorbic acid.
- Equipment: Laminar airflow cabinet, Autoclave, Culture tubes or jars, Scalpel and forceps, Stereomicroscope, Growth chamber

Protocol in Detail

A) Preparation of Media:

Meristem Culture Medium:

- MS basal salts and vitamins.
- Sucrose: 30 g/L.
- BA: 0.5-1.0 mg/L.
- GA_3: 0.1 mg/L.
- IBA or NAA or IAA: 0.01-0.1 mg/L.
- PVP: 0.5 g/L or Ascorbic acid: 100 mg/L.
- Agar: 7 g/L.
- pH adjusted to 5.8 before autoclaving.

B) Procedure:

- Expose collected apple shoots to a temperature of 37 to 40 C for 4 weeks under controlled conditions to suppress viral activity.
- Rinse shoots with 70% ethanol for 40 seconds. Immerse in 2% sodium hypochlorite solution for 15 minutes. Rinse four times with sterile distilled water.
- Under a stereomicroscope, carefully remove leaf primordia to expose the apical dome.

- Excise meristems measuring 0.2 to 0.7 mm, ensuring minimal damage.
- Place excised meristems onto the prepared culture medium in culture tubes or jars. Incubate at 25 C with a 16-hour light/8-hour dark photoperiod.
- Monitor cultures for contamination and growth over 6 weeks. Record survival and proliferation rates, noting the influence of meristem size and season.
- Subject regenerated shoots to DAS-ELISA to confirm the absence of viruses such as ACLSV, ApMV, ASGV, and ASPV (Virus Indexing).
- Transfer virus-free shoots to fresh medium with appropriate PGR concentrations to promote multiplication.
- Induce rooting by transferring shoots to medium supplemented with suitable auxins (e.g., IBA at 0.5 mg/L).

C) Acclimatization:

- Gradually acclimate rooted plantlets to ex vitro conditions by transferring them to pots with sterile soil mix and maintaining high humidity initially.

Potential Problems Encountered During In Vitro Meristem Culture and Their Troubleshooting

- Low survival of excised meristems is often due to mechanical damage or stress during excision, which can be reduced by using a stereomicroscope and fine, sterile tools to carefully excise meristems between 0.2 to 0.7 mm.
- Contamination in early stages of culture typically arises from incomplete surface sterilization, and can be prevented by using 70% ethanol followed by 2% NaOCl with multiple sterile water rinses.
- Browning and necrosis of meristems may result from oxidative stress, which can be mitigated by incorporating antioxidants such as PVP (0.5 g/L) or ascorbic acid (100 mg/L) in the culture medium.
- Slow or no meristem growth often results from inadequate hormonal balance or poor media composition, which can be improved by optimizing BA (0.5 to 1.0 mg/L) and adding low levels of auxins and GA_3 to stimulate development.
- Failure of shoot elongation or vitrification is usually due to excessive cytokinins or improper gelling, and can be resolved by adjusting BA concentration and ensuring proper agar concentration (7 g/L).
- Regenerated shoots testing virus-positive even after culture may be due to improper thermotherapy duration or large meristem size; use smaller meristems and maintain 4-week thermotherapy at 37 to 40°C prior to excision.

- Rooting failure in virus-free shoots often occurs due to residual cytokinin effects, and can be improved by transferring to auxin-rich rooting medium (e.g., IBA 0.5 mg/L) after shoot multiplication.
- Acclimatization losses post-rooting arise from poor humidity management or weak root systems, which can be minimized by gradually reducing humidity and using sterile, well-aerated potting mix.

CHAPTER 17

CELL SUSPENSION CULTURE IN APPLE (*MALUS × DOMESTICA* BORKH.)

Aim

To establish and maintain a viable apple cell suspension culture system for applications in mass propagation, secondary metabolite production, and genetic transformation.

Principle

Cell suspension culture involves growing dispersed plant cells in liquid media under continuous agitation. These cultures are typically initiated by transferring friable callus into liquid medium. In apple, such systems support uniform cell division, allow easy medium manipulation, and are scalable for biotechnological applications such as transformation, metabolite extraction, or protoplast isolation.

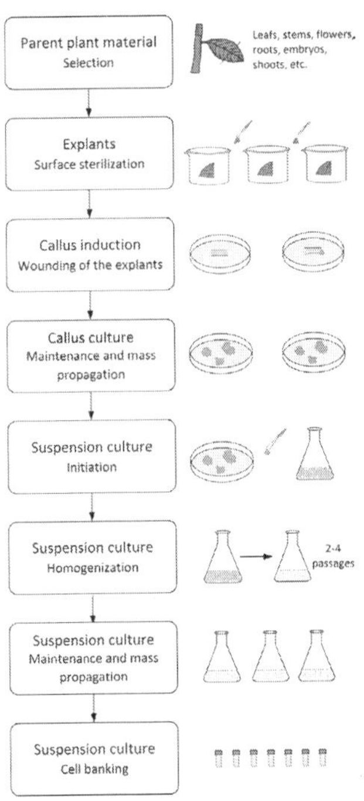

Materials Required

- Apple explants (immature leaves or cotyledons)
- Murashige and Skoog (MS) basal salts
- BAP (6-benzylaminopurine), 2,4-D (2,4-dichlorophenoxyacetic acid), NAA
- Sucrose (30 g/L), Phytagel (for callus induction)

- Culture vessels (Petri dishes, Erlenmeyer flasks, conical tubes)
- Orbital shaker (100 to 120 rpm)
- Sterile pipettes, filter paper, forceps
- pH meter, autoclave, laminar airflow cabinet

Media Composition

Callus Induction Medium (Solid):

- MS basal salts
- 2.0 mg/L 2,4-D + 0.5 mg/L BAP
- 30 g/L sucrose
- 0.25% phytagel

Suspension Culture Medium (Liquid):

- MS basal salts
- 0.1 mg/L 2,4-D + 0.2 mg/L BAP
- 30 g/L sucrose

Protocol in Detail

- Sterilize explants and culture on solid callus induction medium with Adjusted pH to 5.8; autoclave at 121°C for 20 minutes.
- Incubate at 25 C in the dark for 3 to 4 weeks until friable callus forms.
- Transfer ~2 g of friable callus into 50 mL liquid medium in a 250 mL Erlenmeyer flask with Adjusted pH to 5.8; autoclave at 121°C for 20 minutes.
- Incubate on an orbital shaker at 100 to 120 rpm under continuous light or 16/8 h photoperiod.
- Subculture every 7 days by transferring 5 to 10 mL of cell suspension into fresh medium.
- Observe cell clumps under a microscope to ensure uniformity.
- Filter suspension through sterile mesh (300 to 700 μm) if necessary to remove large clumps.
- Measure packed cell volume (PCV) or fresh weight for growth tracking.

Potential Problems Encountered During Suspension Culture and Their Troubleshooting

- Failure to initiate suspension culture is often due to compact or non-friable callus, which can be resolved by using only soft, friable, and pale-yellow callus induced on optimized 2,4-D + BAP solid medium.

- Cell clumping and sedimentation in suspension is commonly caused by inadequate agitation or overgrown tissue, and can be minimized by maintaining 100 to 120 rpm on an orbital shaker and filtering out large clumps using a sterile mesh (300 to 700 µm).
- Low cell viability and poor proliferation may result from inappropriate auxin-cytokinin balance or prolonged subculture intervals, which can be improved by fine-tuning 2,4-D and BAP concentrations and subculturing every 7 days.
- Browning of cells or medium darkening due to phenolic oxidation can be prevented by using antioxidants (like ascorbic acid or activated charcoal) or by frequent medium renewal.
- Foaming or pH fluctuation in the medium may occur due to metabolic activity and agitation, and can be managed by using antifoaming agents (if needed) and ensuring the pH is adjusted to 5.8 before sterilization.
- Bacterial or fungal contamination is typically the result of inadequate aseptic transfer, which can be prevented through strict sterile techniques, filtered air cabinets, and autoclaved tools and vessels.
- Reduced cell growth or early cell death in suspension may be due to depletion of nutrients, which can be addressed by monitoring packed cell volume (PCV) and refreshing the medium before overgrowth.
- Non-uniform growth and asynchronous cell stages can be managed by regular homogenization of culture and using cell sieving steps during subculturing to maintain uniform suspension density.

CHAPTER 18

TRANSFORMATION & REGENERATION IN APPLE (*MALUS* X *DOMESTICA* BORKH.) USING *AGROBACTERIUM*-MEDIATED PROTOCOL

Aim

To achieve stable genetic transformation and regeneration in apple cultivar using *Agrobacterium tumefaciens* and establish virus-free transgenic lines, facilitating genetic improvement and functional genomics studies.

Principle

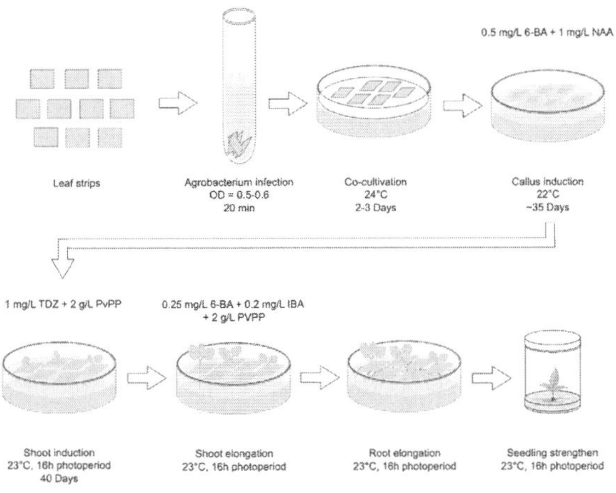

Agrobacterium tumefaciens-mediated transformation exploits the bacterium's natural ability to transfer T-DNA into plant genomes. In apple, leaf explants from in vitro-grown shoots are co-cultivated with *Agrobacterium* carrying the gene of interest. Following selection on antibiotic-containing media, transformed cells regenerate into whole plants through organogenesis.

Materials Required

- In vitro-grown apple shoots
- Bacterial Strain:
- *Agrobacterium tumefaciens* strain (e.g., LBA4404) with target plasmid
- Murashige and Skoog (MS) and N6 media salts and vitamins
- 70% ethanol, 1% sodium hypochlorite
- Phytagel, Bacto agar
- Sucrose
- Plant growth regulators: 6-Benzylaminopurine (BAP), Indole-3-butyric acid (IBA), Naphthalene Acetic Acid (NAA)
- Antibiotics: Kanamycin (25 mg/L), Meropenem (50 mg/L)
- Acetosyringone (20 µM)
- Equipment: Laminar airflow cabinet, Autoclave, Growth chamber, Scalpel and forceps, Petri dishes and culture vessels, Sonicator
- PCR equipment for verification

Protocol in Detail

A) Preparation of Media:

Co-incubation Medium:

- N6-macro
- MS-micro
- 0.5% MS vitamins
- 5 mg/L BAP + 0.2 mg/L NAA
- 20 µM Acetosyringone
- 3% sorbitol
- 0.2% phytagel (pH 5.6)

Regeneration Medium:

- N6-macro
- MS-micro

- 0.5% MS vitamins
- 5 mg/L BAP
- 0.2 mg/L NAA
- 3% sorbitol
- 25 mg/L kanamycin
- 50 mg/L meropenem
- 0.25% phytagel (pH 5.6)

Propagation Medium:

- MS
- 0.2 mg/L BAP
- 0.1 mg/L IBA
- 3% sucrose
- 0.7% Bacto agar (pH 5.8)

Propagation Medium (Transformed):

- MS
- 1 mg/L BAP
- 0.1 mg/L IBA
- 3% sucrose
- 25 mg/L kanamycin
- 50 mg/L meropenem
- 0.7% Bacto agar (pH 5.8)

Rooting Solution:

- 0.5x MS
- 0.1 mg/L IBA (no sorbitol) (pH 5.8)

B) Procedure:

Sterilization & Culture Initiation:

- Harvest shoot tips, trim to 2 to 3 cm, dip in 70% ethanol for few seconds, then 1% sodium hypochlorite for 3 minutes.
- Rinse three times with sterile distilled water.
- Place in 0.5x MS medium with 0.1 mg/L BAP, 0.65% agar, and 3% sucrose. Grow at 23 C under 16/8 h light/dark.

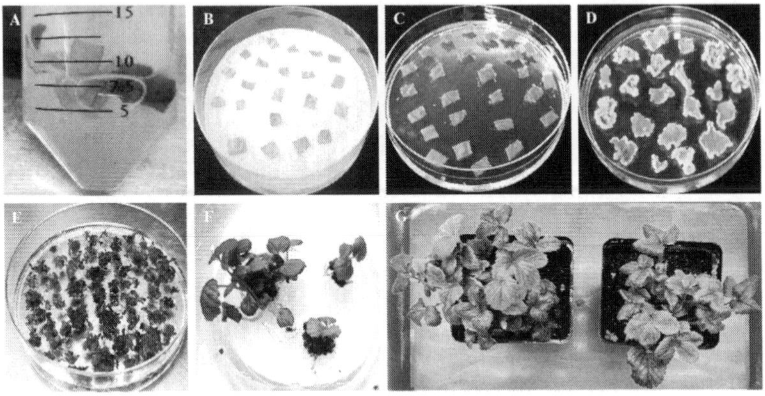

Leaflet Preparation:

Select unfolding leaflets from apex.

Transfer ~500 leaflets onto fresh 0.5x MS medium for pre-conditioning.

Agrobacterium Preparation:

- Grow LBA4404 in ψB medium with 50 mg/L kanamycin at 28°C (OD600 0.8 to 1.2).
- Centrifuge, wash twice in 0.5x MS, resuspend in 20 mL same medium.

Transformation:

- Dip leaflets in *Agrobacterium* suspension and sonicate for 10 seconds at 37 kHz.
- Incubate in suspension for 15 minutes.
- Dry on filter paper, then place on co-incubation medium. Keep in dark at 20 to 23°C for 7 days.

Regeneration:

- Cut leaflets vertically to veins and fragment (2 to 3 per leaflet).
- Place ~20 fragments per plate on regeneration medium. Incubate in dark at 20 to 23°C.
- After 1 month, transfer to light and subculture monthly.
- After ~3 months, green shoots should appear.

Propagation of Transformants:

- Transfer regenerated shoots to transformed propagation medium with antibiotics.

Elimination of Agrobacterium:

- Chop 3 leaflets from regenerants in sterile water, shake for 1 hour.

- Plate supernatant on LB agar with 50 mg/L kanamycin.
- Incubate at 28°C for 2 to 3 weeks. If colonies appear, confirm by plasmid-specific PCR.

Rooting and Acclimatization:

- Transfer shoots to rooting solution in perlite: vermiculite (1:1) mix.
- Cover with plastic bag. Once rooted, gradually acclimate by perforating the bag.

Potential Problems Encountered During Transformation & regeneration and Their Troubleshooting

- Low transformation efficiency is often caused by suboptimal *Agrobacterium* density or infection parameters, which can be improved by maintaining OD between 0.8 to 1.2, adding Acetosyringone (20 μM), and optimizing sonication and co-cultivation conditions.
- Necrosis of explants post-co-cultivation typically results from *Agrobacterium* overgrowth, which can be mitigated by reducing co-incubation time, using meropenem (50 mg/L), and drying explants before transfer to solid medium.
- Bacterial contamination in regenerated shoots occurs due to incomplete elimination of *Agrobacterium*, and can be resolved by regular subculturing onto antibiotic-supplemented media and confirming sterility with LB plating and PCR.
- Failure of shoot regeneration from leaf explants may stem from incorrect hormone balance or excessive tissue stress, which can be addressed by using optimized regeneration media (5 mg/L BAP + 0.2 mg/L NAA) and maintaining initial darkness.
- Chimeric or partially transformed shoots arise from transient expression and can be avoided by early selection pressure using kanamycin (25 mg/L) and subculturing onto fresh selective media at regular intervals.
- Poor rooting of transformed shoots may result from residual cytokinin effects or high salt concentration, which can be corrected by transferring to half-strength MS medium with low IBA (0.1 mg/L) and omitting sorbitol.
- Delayed or reduced shoot elongation in transformed lines may be due to excessive antibiotic stress, which can be minimized by fine-tuning kanamycin and meropenem concentrations and alternating with non-selective medium when necessary.

- High mortality during acclimatization often occurs due to weak root systems or sudden humidity loss, and can be prevented by using a gradual hardening process in perlite: vermiculite mix under humidity domes.

CHAPTER 19

TEMPORARY IMMERSION BIOREACTOR (TIB) SYSTEM FOR LARGE-SCALE APPLE MICROPROPAGATION

Aim

To establish a scalable and efficient protocol for the mass multiplication of apple plantlets using a Temporary Immersion Bioreactor (TIB) system.

Principle

The TIB system alternates between immersing plant tissues in liquid nutrient media and exposing them to air. This intermittent immersion enhances nutrient uptake and gas exchange while minimizing issues like Hyperhydricity, common in continuous liquid cultures. The method is particularly effective for woody plants like apple, facilitating uniform shoot proliferation and rooting.

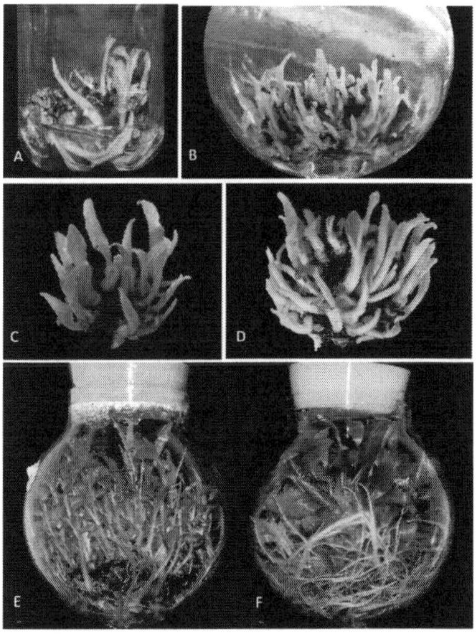

Materials Required

- Explant Source: Actively growing shoot tips or nodal segments (1.0 to 1.5 cm) from virus-free apple cultivars (e.g., M.9 EMLA, M.26).

- Bioreactor Setup: RITA® or equivalent TIB units with dual chambers for medium and explants.
- Culture Media: Murashige and Skoog (MS) basal medium.
- Plant Growth Regulators (PGRs):
- 6-Benzylaminopurine (BAP)
- Indole-3-butyric acid (IBA)
- Additives: Sucrose
- Agar (for semi-solid comparisons)
- pH adjustment reagents (NaOH, HCl)
- Sterilization Equipment: Autoclave, laminar airflow hood.
- Culture Room Conditions: Temperature: 25 ± 2 C; Light: 16-hour photoperiod with 3000 lux intensity

Media Composition

Multiplication Medium:

- MS basal salts
- Sucrose: 30 g/L
- BAP: 7.5 mg/L
- pH: Adjust to 5.8 before autoclaving

Rooting Medium:

- MS basal salts
- Sucrose: 30 g/L
- IBA: 0.5 mg/L
- pH: Adjust to 5.8 before autoclaving

Detailed Protocol

1. Explant Preparation

- Excise shoot tips or nodal segments (1.0–1.5 cm) from healthy, disease-free apple plants.
- Surface sterilize explants using standard protocols (e.g., 70% ethanol for 30 seconds, followed by 0.1% mercuric chloride for 5 minutes), then rinse thoroughly with sterile distilled water.

2. Initiation Phase

- Inoculate sterilized explants onto semi-solid MS medium supplemented with 1.0 mg/L BAP to initiate shoot proliferation.

- Incubate under controlled conditions (25 ± 2 °C, 16-hour photoperiod) for 2 to 3 weeks until shoots develop.

3. Transfer to TIB System

- Assemble the TIB units, ensuring sterility.
- Add 250 mL of multiplication medium to the lower chamber of each bioreactor.
- Place 5 explants into the upper chamber of each TIB unit.

4. Immersion Parameters

- Set the immersion frequency to 2 minutes every 6 hours.
- Maintain cultures under the same controlled environmental conditions.

5. Multiplication Phase

- Cultivate explants in the TIB system for 4 weeks.
- Monitor for shoot proliferation, ensuring shoots are healthy and free from Hyperhydricity.

6. Rooting Phase

- Transfer proliferated shoots to fresh TIB units containing rooting medium.
- Maintain the same immersion parameters and environmental conditions.
- Cultivate for an additional 3 to 4 weeks until adequate root systems develop.

7. Acclimatization

- Carefully remove plantlets from the bioreactor.
- Rinse roots gently to remove any residual medium.
- Transplant plantlets into pots containing a sterilized substrate mix (e.g., peat moss: vermiculite, 1:1).
- Gradually acclimate plants to ambient conditions by reducing humidity over 2 to 3 weeks.

Potential Problems Encountered During Temporary Immersion Bioreactor (TIB) System Culture and Their Troubleshooting

- Each problem commonly faced during large-scale micropropagation in a TIB system is paired with a concise, scientifically sound solution:
- Hyperhydricity of shoots, often due to excessive immersion duration or frequency, can be minimized by reducing immersion time (e.g., 2 minutes every 6 hours) and improving ventilation in the bioreactor design.

- Contamination in the bioreactor system usually arises from poor sterilization or inadequate sealing and can be prevented by thoroughly sterilizing all components and ensuring aseptic assembly inside a laminar airflow cabinet.
- Non-uniform shoot proliferation may result from inconsistent nutrient exposure, which can be corrected by standardizing explant size and placement, and optimizing liquid media volume per explant load.
- Shoot tip necrosis can occur due to low humidity or mechanical stress and can be alleviated by maintaining proper humidity in the culture room and avoiding physical damage during handling.
- Poor rooting response is often linked to carryover effects of high cytokinins from multiplication media, which can be managed by transferring shoots to IBA-enriched rooting medium with no BAP carryover.
- Foaming or frothing in the bioreactor, particularly during longer culture periods, may result from high organic loads and can be reduced by using antifoaming agents or regular medium replacement.
- Reduced shoot elongation in TIBs may occur due to light limitation or hormonal imbalance, and can be improved by supplementing elongation media with GA_3 or slightly lowering cytokinin concentration.
- Acclimatization losses post-biotransfer are commonly due to residual moisture and weak cuticle formation, which can be mitigated by careful rinsing, gradual humidity reduction, and use of well-aerated substrates.

CHAPTER 20

IN VITRO MICROGRAFTING OF APPLE (*MALUS* × DOMESTICA *BORKH.*) *SCIONS*

Aim

To develop a reliable in vitro micrografting protocol for apple scions onto in vitro-grown rootstocks, facilitating rejuvenation, virus elimination, compatibility studies, and rapid multiplication of elite genotypes.

Principle

In vitro micrografting involves the precise union of a small scion (usually a shoot tip or meristem from a donor plant) onto a juvenile or dwarf rootstock under aseptic conditions. This technique is particularly valuable in apple tissue culture for rejuvenating mature scions, overcoming graft incompatibility, and eliminating viruses or systemic pathogens. Unlike conventional grafting, micrografting is performed under sterile conditions, often using rootstocks and scions that have been propagated or cleaned via tissue culture. Successful union leads to the formation of a vascular connection, allowing the scion to develop into a complete, healthy plant with the desired genetic traits of the donor.

Materials Required

- In vitro cultured apple rootstocks (e.g., M9 or M26 clones)
- Donor material: apical shoot tips or meristems of elite apple cultivars
- Murashige and Skoog (MS) basal medium
- Plant Growth Regulators (optional):

- BAP (0.1 - 0.5 mg/L)
- GA_3 (0.1 - 0.2 mg/L)
- IBA (0.2 mg/L) for rooting post-graft
- Sucrose (3%)
- Agar (0.7%)
- Forceps, sterile scalpel or microblade, micrografting board or platform
- Growth chamber, Sterile laminar airflow cabinet
- Transparent parafilm, strips of sterile cling film, or sterile silicon sleeves (optional)

Media Composition

Rootstock Multiplication Medium:

- MS basal salts + vitamins
- Sucrose - 30 g/L
- BAP - 0.5 mg/L (optional)
- Agar - 7 g/L

Post-Graft Establishment Medium:

- MS basal salts
- GA_3 - 0.1 mg/L
- BAP - 0.2 mg/L
- Sucrose – 20 - 30 g/L
- Agar - 7 g/L

Protocol in detail:

- Establish and multiply rootstocks in vitro using MS medium.
- Choose vigorous, straight shoots of ~2.5 to 3 cm length with a minimum stem thickness of 1 to 1.5 mm.
- Maintain rootstocks in upright position for 1 week before grafting.
- Excise shoot tips (scions) of ~0.3 to 0.5 cm from elite apple donor cultivars using a dissecting microscope to ensure precision.
- Surface sterilize scion donor tissue (if sourced ex vitro) using 70% ethanol (30 sec) followed by 1% NaOCl (10 min), and rinse with sterile distilled water (3).
- Perform the grafting inside a laminar airflow cabinet under sterile conditions.

- Using a sterile microblade, decapitate the rootstock just above the first node to create a clean vertical surface.
- Prepare a vertical slit (3 to 4 mm deep) in the center of the cut surface of the rootstock (side slit or wedge cut can also be used depending on rootstock thickness).
- Prepare a tapered scion base to fit snugly into the slit of the rootstock.
- Insert the scion gently into the slit so that vascular continuity is maximized.
- Optionally wrap the union site with a sterile thin strip of parafilm or cling film for support.
- Transfer the grafted plantlets to post-graft establishment medium with adjusted pH to 5.8.
- Incubate grafted cultures under low light (8 to 10 µmol m^{-} s^{-}) for 3–5 days to minimize desiccation.
- Gradually increase light intensity to standard levels (40 to 50 µmol m^{-} s^{-}).
- Maintain at 25 ± 2 C under 16/8 h light/dark photoperiod.
- Remove parafilm (if used) after 10 to 12 days once the graft union is stabilized.
- Shoots should elongate after 2 to 3 weeks if grafting is successful.
- Once shoots reach 3 to 4 cm, transfer to rooting medium (½ MS + IBA 0.5 mg/L) with adjusted pH to 5.8.

Acclimatization of Grafted Plantlets:

- Carefully remove rooted grafted plantlets from culture vessels and wash off agar residues.
- Transfer to pots containing sterilized potting mix (peat: vermiculite: sand 1:1:1).
- Maintain in a high humidity environment under polyethylene covers or mist chambers.
- Gradually harden over 3–4 weeks by progressively exposing to ambient humidity.

Potential Problems Encountered during In Vitro Micrografting and their Troubleshooting

- Each problem is merged with its corresponding solution in a precise, expert tone that fits seamlessly with the rest of your chapters:
- Failure of graft union formation is often due to poor scion-rootstock alignment or loose insertion, which can be resolved by precisely matching vascular tissues and securing the graft interface with sterile parafilm or a silicon sleeve.

- Desiccation of the scion tip after grafting is commonly caused by high light intensity or low humidity in early stages and can be minimized by maintaining low light (8–10 µmol m^{-} s^{-}) for 3–5 days and covering with transparent lids or cling wrap.
- Scion necrosis or browning may result from oxidative stress or damage during excision, which can be prevented by using freshly cut, juvenile shoot tips and handling them gently under sterile conditions.
- Rootstock collapse or contamination after grafting is often due to pre-existing microbial load or poor media handling, which can be addressed by ensuring complete surface sterilization and using freshly prepared sterile media.
- Callus formation without successful shoot elongation may result from hormonal imbalance at the graft interface, which can be corrected by adjusting GA_3 and BAP concentrations in the post-graft establishment medium.
- Delayed graft success or failure under thicker or lignified rootstocks can be avoided by selecting young, flexible rootstock tissue (1–1.5 mm diameter) and using a wedge or omega cut to increase contact surface.
- Failure of root induction in elongated grafted shoots may occur due to insufficient auxin exposure or poor physiological state, which can be resolved by transferring shoots to ½ MS medium supplemented with 0.5 mg/L IBA.
- High mortality during acclimatization is typically due to abrupt humidity changes or weak root systems and can be prevented by gradual hardening and use of well-drained, sterile potting substrate under high-humidity chambers.

CHAPTER 21

EMERGING FRONTIERS IN *IN VITRO* PLANT BIOTECHNOLOGY: NANOMATERIALS, BIOSENSORS, AND BIOTIZATION

Use of Nanotechnology in Plant Tissue Culture

Introduction

Nanotechnology is the study of managing minute particles to make new materials or modifying existing ones. Professor Norio Taniguchi at the University of Tokyo in 1974 first coined the term nanotechnology. It is defined as an innovation to obtain the additional high precision and ultra fine measurements, for example the exactness and fineness in the range of 1 nanometer (nm or 10^{-9} m long). The term Nano of nanomaterials was originated from the Greek word signifying "dwarf". The word nano implies 10^{9} or one billionth of a meter and is utilized for materials with a size extending somewhere in the range of 1–100 nm. Nanomaterials are usually having very specific and special characters owing to their very small size which also increases their surface area, which in turn provides more interactive surface. In plant tissue culture, nanoparticles can be used either as nutrients or as disinfectant or both.

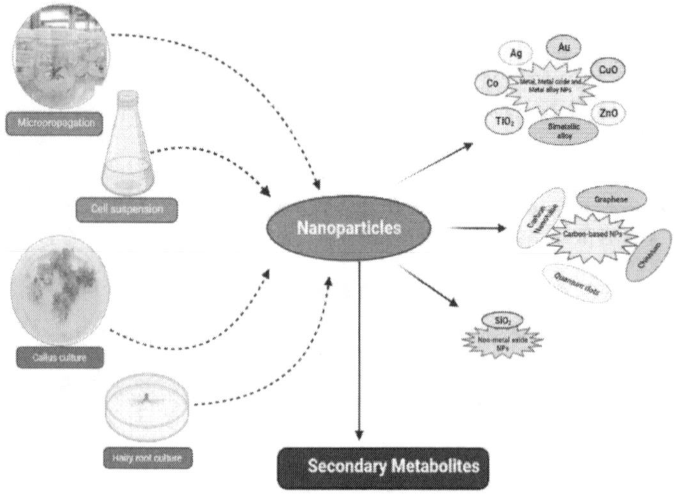

Nano-Particles for Surface Sterilization of Explants

In tissue culture, microbial infections, both bacterial and fungal is a significant issue. Surface cleansing of the explants is pre-requisite for in-vitro culture inception since microorganisms develop quicker in culture media than the explants, which are capable of culture commencement. Generally, for the establishment of in vitro experiments, explants are taken from fields. These explants are then surface sterilized using different chemicals. Commonly used disinfectants in tissue culture include ethanol, bromine water (BW), hydrogen peroxide (H_2O_2), mercuric chloride ($HgCl_2$), sodium hypochlorite (NaOC1), calcium hypochlorite (CaOCl), silver nitrate ($AgNO_3$), fungicides and anti-infection agent. The use of higher concentration of disinfectants may damage the explants and hence, appropriate concentration and duration should be standardized. Several metal and metal oxide nanoparticles possess antibacterial properties and most common nanoparticles used as disinfectants include ZnO, Ag and TiO_2. Other nanoparticles such as ZnO, Fe_2O_3, MgO, Al_2O_3, CuO, Ni, SiO_2, TiO_2 etc. are used in plant tissue culture as sanitizing agents for explants and do not have any negative effect on explants. Though, seed germination, cotyledon and leaf endurance might be sometimes diminished if explants are treated with higher levels of silver nanoparticles. Removal of contaminants in plant tissue culture depends upon the type, size and dispersion of the nanoparticles. The endurance and recovery of explants can be increased with the application of nanoparticles. Hence, optimizing nanoparticle concentration for different plant species is important to sterilize the targeted explant without any phyto-toxicity. The combined impact of sanitizing agents or antimicrobials and NPs can improve the viability of explants.

Surface Sterilization Protocol for Apple Using Nano-Particles

1. Select healthy leaves of strawberry (explant). Wash them twice with sterile distilled water.

2. Wash the leaf explants with 70% ethanol for 2 minutes followed by one wash with distilled water.
3. Add sodium hypochlorite (4% active chlorine) and a drop of tween 80 to the leaves and wash them with continuous shaking for 15 mins.
4. Wash the leaf explants with sterile distilled water for 2-3 times.
5. Treat the leaf explants with 200 mg/L AgNPs solution for 20 min followed by thorough washing (4-5 times) with sterile distilled water.
6. Place the surface sterilized leaf explants in an appropriate media for germination or callus.
7. Wrap the junction of the petri dishes and the test tube with Parafilm.
8. Place the cultures on the culture shelf at appropriate temperature and light for growth and incubate at appropriate temperature in dark for callus induction.

Role of Nanoparticles in Callus Induction, Shoot and Root Growth

Nanoparticles play an important role in callus formation, root and shoot generation. Addition of silver nanoparticles has been shown to increase the percentage of shoot induction, number of shoots and formation of callus. Delayed senescence and enhanced survival rate by down regulation of TuACS gene was also observed by adding AgNPs. Positive effects of silver nanoparticles on organogenesis might be attributed to hindrance of ethylene creation. Zinc oxide nanoparticles (ZnO NPs) also have positive effects on callus proliferation. Adding nanoparticles to tissue culture medium influences proliferation of callus, multiplication of shoots, somatic embryogenesis and formation of roots by changing the activity of antioxidant enzymes, expression of genes and hindrance in ethylene generation. Metal and metal oxide nanoparticles increase the morphogenetic capacity of explants.

Callus Induction Protocol for Apple Using Nano-Particles

1. 1. Select healthy leaves of Apple (explant). Wash them twice with sterile distilled water.
2. Wash the leaf explants with 70% ethanol for 2 minutes followed by one wash with distilled water.
3. Add sodium hypochlorite (4% active chlorine) and a drop of tween 80 to the leaves and wash them with continuous shaking for 15 mins.
4. Wash the leaf explants with sterile distilled water for 2-3 times.
5. Treat the leaf explants with 200 mg/L AgNPs solution for 20 min followed by thorough washing (4-5 times) with sterile distilled water.
6. Place the surface sterilized leaf explants in callus induction media (CIM) composed of MS media + 3% sucrose + 0.8% agar +1 mg L^{-1} NAA + 2 mg L^{-1} BA + 10 mg L-1 ZnO NPs

7. Wrap the junction of the petri dishes and the test tube with Parafilm.
8. Incubate the petri dishes for 21 days in darkness at 28°C in an incubator for callus induction followed by transfer to culture shelf at 24 ± 2 °C with 16/8 h photo period light for growth.
9. Subculture the calli on a fresh CIM at 2-week interval

Effect of Nano-Materials on Genetic Transformation

Common methods of gene transfer to the plant cells, tissues and organs are *Agrobacterium* mediated transformation, electroporation and particle bombardment. Many tissue culture products are utilized as source of plant tissue like callus cultures, somatic embryos or protoplasts. Generally, electroporation is utilized to transfer genes into the protoplasts, however isolation and regeneration of chloroplast is difficult. Nanoparticles also find a number of applications to ease the genetic transformation in plant tissue cultures. Nanoparticles are used for protoplast isolation to minimize the effects of cellular enzymes. DNA is transferred into the tobacco protoplast with the help of mesoporous silica nanoparticles by endocytosis. Similarly, biolistic gun is used to transfer gold coated silica nanoparticles, chemicals and DNA to the callus and leaves which result in lower plant cell injuries and significant recovery of plant tissue. Poly-(amidoamine) dendrimer nanoparticles is an innovative technique which can efficiently transfer green-fluorescent protein (GFP)-encoding plasmid DNA into turf-grass cells. The addition of silver nanoparticles in culture media effectively improves the morphology and anatomy of the callus by modifying the DNA and protein profile in the calli.

Nanomaterials for Elicitation of Bioactive Compounds

Plants produce various bioactive compounds or secondary metabolites that can play an essential role in health and even survival in response to different stresses. Plant tissue culture can provide economical, viable and sustainable production of different secondary metabolites such as alkaloids, terpenoids, quinones, steroids, and phenylpropanoids. In vitro plant cell and organ cultures are highly beneficial for generating bioactive compounds. By optimizing culture conditions and the culture media composition, and by supplementing the precursors and elicitors, the production level of these compounds can be significantly enhanced. In a culture medium, the added NPs may act as an elicitor and as a nutrient source. Nanomaterials including single-walled carbon nanotubes (SWCNTs), multiple walled carbon nanotubes (MWCNTs), fullerenes and chitosan nanoparticles have been applied to different medicinal plant species to study their impact on secondary metabolite production. Callus culture using MS medium supplemented with 10 mg/L CuO NPs increases the production of bioactive compounds like phenolics and flavonoids, rosmarinic acid, chicoric acid, and eugenol in *Occimum basilicum*.

Use of Biosensors in Plant Tissue Culture

A biosensor is an integrated receptor–transducer device that can convert a biological response into an electrical signal, biosensors for a wide range of applications such as health care and disease diagnosis, environmental monitoring, water and food quality monitoring, and drug delivery. The engineering of biosensors provides a new tool for the monitoring of phytohormones in plant tissue culture media. The visualization and quantification of phytohormone distribution can be realized by using biosensors and other related technologies including fluorescent sensors, transcriptional reporters, degradation sensors, and luciferase. Fluorescence biosensors are particularly effective for monitoring the spatial and temporal distribution of small signal molecules, which can be used to study the distribution and content of plant hormones in the subcellular.

The standardization of a medium formula, the mineral and hormone composition of the plant micropropagation medium is an important factor in the growth of explants, but the different plant species lead to different nutritional and hormone requirements. The development and application of a predictive modeling system for medium formulation and culture conditions can improve the efficiency of culture optimization. Data-driven modeling is considered as an effective alternative to optimizing biological processes and nonlinear multivariate modeling. Modeling systems are often combined with optimization algorithms to improve efficiency. Appropriate modeling techniques can not only predict the value of the sensor, but also estimate the probability. Commonly used modeling technology systems include the combination of an artificial intelligence (AI) model and optimization algorithm (OA) and the combination of gene expression programming (GEP) and genetic algorithm (GA).

Use of Plant Growth Promoting Microbes and Mycorrhiza in Plant Tissue Culture

Plant Growth Promoting Microbes (PGPMs) are selected microorganisms usually found in the soil and roots of plants (rhizosphere) which have positive impact on plant growth, development and survival. Plant Growth Promoting Microbes include Plant Growth-Promoting Rhizobacteria (PGPR) and Plant Growth-Promoting Fungi (PGPF). Plant Growth Promoting Microbes are involved in the degradation, mineralization of organic compounds, solubilization of inorganic compounds and release of biologically active compounds like chelators, phytohormones, and antibiotics, among others, which are necessary for the general well-being of plants. The use of competent PGPMs in tissue culture under in vitro and ex vitro conditions is called "biotization". Microplant biotization is a biotechnological practice aimed at reducing chemical input in plant production. The biotization can be done at all stages of in vitro propagation namely; Stage 0 (Plant stock immobilization and pre-treatments, selection of the explants), Stage I (Culture establishment), Stage II (Elongation and multiplication), Stage III (Rooting), Stage IV (Weaning, hardening, and acclimatization) and Stage V (Transfer under natural conditions-to the field). In stages II and III of micropropagation by micro-cutting, PGPMs act generally as bio-stimulants by promoting elongation and increasing rooting, while in stage IV, they act as biocontrol agents and help to deal with biotic and abiotic stress factors. In addition to their three main growth promoting mechanisms namely; direct, indirect/direct and indirect, certain PGPMs such as Rhizobium, Frankia, Bradyrhizobium, and mycorrhizal fungi have been recognized to be able to improve the physical properties of the soil by making it more conducive for plant growth and development.

Beneficial Effects of Plant Growth Promoting Fungi in In Vitro Plant Culture

Plantlets obtained from micropropagation through tissue culture are often adversely affected by water stress; because of low water absorption capacity of their roots. Inoculation with Arbuscular Mycorrhizal Fungi (AMF) in vitro is an important tool to deal with this problem. Through biosynthesis of phytohormones or Plant Growth Regulators (PGRs), AMF impact on post-transplant performance of in vitro grown plants by increasing nutrients availability and inducing resistance to pathogens. Fungi produce phytohormones such as auxins, cytokinins (CKs), abscisic acid (ABA), gibberellic acids (GAs), ethylene (ET), salicylic acid (SA), and jasmonic acid (JA). These hormones control plant development and activate signaling pathways during biotic and abiotic stress. Arbuscular Mycorrhizal Fungi also increase the uptake of micronutrients such as P, Zn, Cu, Fe, etc. and plays an important role during the acclimatization phase, because the adventitious and weak root system, without root hair, of in vitro plants do not allow optimal absorption of nutrients from the soil during the early stage of the weaning step. Arbuscular Mycorrhizal Fungi can help to overcome this problem, through their arbuscules and hyphae which transfer nutrients, especially phosphate from the soil to the plant.

Beneficial Effects of Plant Growth Promoting Bacteria in In Vitro Plant Culture

Biotized plant tissue cultures benefit from microbial presence through an improvement in photosynthetic efficiency and biomass production. Generally, PGPB improves growth by releasing PGRs required for *in vitro* propagation. Auxins and cytokinins biosynthesis are widespread among rhizobacteria. Many bacteria can produce cytokinins in pure culture and more than 80% of soil bacteria in the rhizosphere can produce auxins especially indole-3-acetic acid (IAA), which is the major auxin active form in plants. Plant Growth Regulators from microorganisms play a compensatory role, especially when micro propagated plants are under sub-optimal environment with insufficient endogenous production. Some plant growth promoting rhizobacteria (PGPR) can induce the production of phytohormones by the plant. Bacterial phytohormones such as gibberellins can interact with other hormones to support elongation. Other phyto regulators such as abscisic acid and salicylic acid are produced by PGPB and can regulate both growth and senescence by modulating ethylene levels in the plant tissue which plays an essential role in the plant defense mechanisms against infections.

Bacteria can produce volatile metabolites which can induce organogenésis improve the efficiency of photosynthesis and provide protection against abiotic stressors. Plant Growth Promoting bacteria and fungi living in the rhizosphere induce systemic resistance (ISR) and enhance defense against a broad range of pathogens and insects. Some PGPB (e.g., Pseudomonas and Bacillus) and some PGPF (e.g., Trichoderma) can sensitize the plant immune system for enhanced defense without directly activating costly defenses.

CHAPTER 21

STRATEGIC TROUBLESHOOTING IN PLANT TISSUE CULTURE: GUIDE FOR COMMON IN VITRO CULTURE PROBLEMS

Introduction

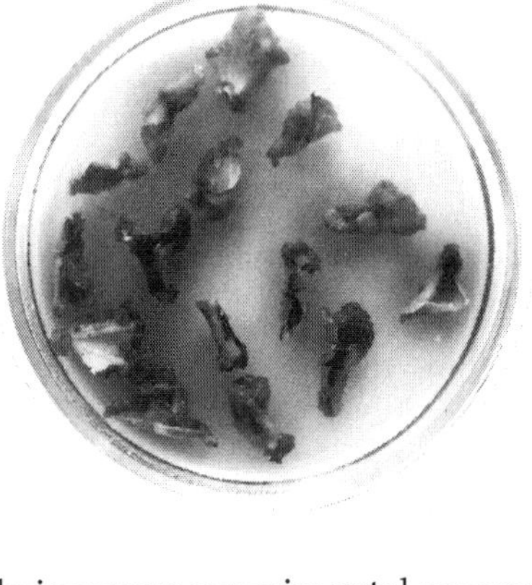

Plant tissue culture, while offering immense potential for mass propagation, genetic improvement, and metabolite production, is highly sensitive to a range of technical and physiological challenges. Even with optimized protocols, practitioners often encounter recurring issues that compromise the health and viability of cultures. This chapter brings together the most common in vitro culture problems across different stages i.e; initiation, maintenance, multiplication, and regeneration and provides scientifically validated, experience-based troubleshooting strategies to overcome them. These insights not only improve experimental success but also ensure reproducibility and efficiency in research and commercial settings.

Browning of Explants

One of the most frequent and visually distressing issues in tissue culture is browning of explants and the surrounding medium, usually observed within the first few days of culture. This browning is primarily caused by the oxidative polymerization of phenolic compounds released from wounded tissues, catalyzed by polyphenol oxidases (PPOs). Such oxidative stress not only darkens the tissues but often leads to complete necrosis of the explant.

To counter this, researchers should pre-treat explants with antioxidant solutions such as ascorbic acid or citric acid (typically at 100 mg/L) to neutralize phenolic exudates. Incorporating activated charcoal (0.1 to 0.5%) in the culture medium also helps by adsorbing these toxic compounds. Rapid transfer of explants to media post-excision and minimizing handling time can further reduce oxidation. Frequent subculturing to fresh media can also dilute the toxic effects and help sustain explant viability during the critical establishment phase.

Hyperhydration (Vitrification)

Hyperhydration, or vitrification, manifests as translucent, water-soaked, and often brittle shoots with distorted morphology. This physiological disorder is primarily triggered by an imbalance in cytokinin concentration, especially when using highly active types like thidiazuron (TDZ) or excessive BAP, combined with high relative humidity in culture vessels.

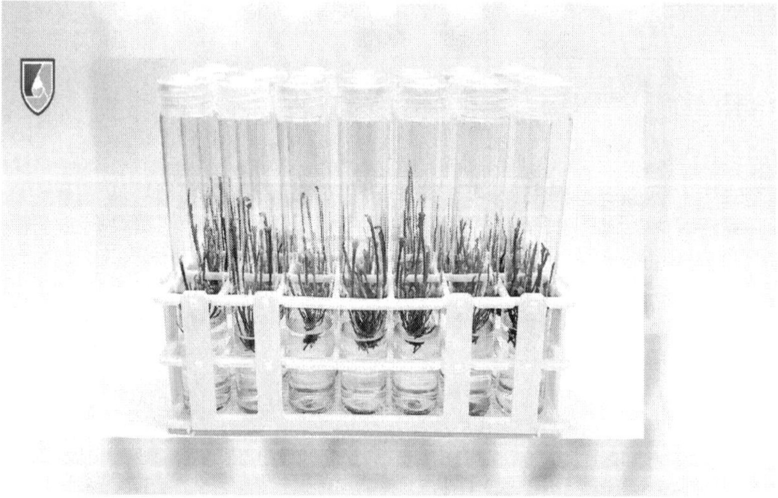

To mitigate vitrification, it is advisable to reduce cytokinin concentrations to physiological levels or switch to milder alternatives such as zeatin. Increasing the concentration of gelling agents like agar (to 0.8%) or replacing agar with phytagel can provide better texture to the medium and reduce excess water availability. Additionally, using culture vessels with better gas exchange or breathable closures, such as cotton plugs or parafilm with micro-punctures, helps maintain optimal humidity around the cultures and reduces vitrification.

Contamination Control

Microbial contamination remains one of the most pervasive and frustrating problems in plant tissue culture laboratories. Bacteria, fungi, or yeasts can quickly colonize culture media, often within 3 to 7 days of inoculation, resulting in tissue collapse and experimental failure. Contamination sources may include improperly sterilized explants, unclean instruments, airborne spores, or even endophytic microorganisms residing within plant tissues.

Effective control begins with a rigorous surface sterilization protocol using a combination of 70% ethanol and sodium hypochlorite (1 to 2%) along with a surfactant like Tween-20 to break surface tension. Each sterilant should be followed by multiple rinses with sterile distilled water. For particularly recalcitrant materials or woody tissues, pre-culture in antibiotic-supplemented media (e.g., with cefotaxime or streptomycin) may be used. It is equally crucial to maintain sterile conditions during handling autoclave all instruments, work under a laminar airflow hood, and flame sterilize tools between samples. Consistent sanitation routines significantly reduce the microbial load and ensure longer-lasting cultures.

Shoot and Rooting Failure

In many cases, cultures remain green and apparently healthy yet fail to produce shoots or roots even after several weeks on induction media. This issue is often linked to an inappropriate hormonal balance, poor-quality plant growth regulators (PGRs), or the inherent physiological dormancy of the explants. The ratio of cytokinins to auxins must be carefully modulated: higher cytokinin concentrations (e.g., BAP, TDZ) favor shoot induction, while elevated auxins (e.g., IBA, IAA) promote root formation.

To resolve shoot or rooting failure, one should consider altering the cytokinin-to-auxin ratio based on the plant species and explant type. Switching the type of PGR (e.g., replacing BAP with kinetin or IBA with NAA) can often make a substantial difference. Additionally, using pulse treatments short-duration exposure to high auxin concentration has shown success in stimulating root initiation in some woody species. It's also essential to ensure the freshness and proper storage of hormone stock solutions to avoid degradation and reduced bioactivity.

General Culture Degeneration and Long-Term Decline

Cultures that initially perform well may gradually lose vigour over time, exhibit necrosis, or fail to regenerate effectively during subculturing. This decline often results from prolonged exposure to the same medium, accumulation of metabolic by-products, nutrient depletion, or loss of totipotency due to genetic or epigenetic changes.

To combat long-term degeneration, subculturing should be performed every 2 to 4 weeks, depending on the growth rate and type of tissue. Nutrient-rich media should be refreshed regularly, and pH should be maintained around 5.7 to 5.8 for most species. Using additives such as glutamine, casein hydrolysate, or activated charcoal can enhance metabolic support and detoxify the environment. Avoid over-subculturing the same lines beyond 6 to 8 cycles to reduce somaclonal variation or chromosomal instability. Maintaining a record of passage history is crucial for reproducibility and quality control.

GLOSSARY OF KEY TERMS

- **2,4-D (2,4-Dichlorophenoxyacetic acid):** A synthetic auxin used extensively in tissue culture to induce callus formation and cell division.
- **Acclimatization:** The gradual adaptation of in vitro-grown plantlets to ex vitro conditions such as lower humidity and natural light.
- **Agar:** A gelatinous substance derived from red algae, commonly used as a gelling agent in solid plant tissue culture media.
- **Anther Culture:** A technique to induce haploid embryogenesis from pollen-containing anthers under in vitro conditions.
- **Aseptic Technique:** Sterile procedures used to prevent contamination in plant tissue culture, typically performed inside a laminar airflow cabinet.
- **Auxins:** A class of plant hormones (e.g., IAA, NAA) that regulate cell elongation, root formation, and tissue differentiation.
- **BAP (6-Benzylaminopurine):** A synthetic cytokinin used to stimulate shoot formation and direct regeneration in explants.
- **Bioreactor:** A controlled system for culturing cells or tissues on a large scale, such as TIB or RITA® systems.
- **Callus:** A mass of unorganized plant cells induced from explants in vitro; a key intermediate in regeneration protocols.
- **Colchicine:** An alkaloid chemical used for chromosome doubling in haploid cells to produce fertile doubled haploids.
- **Cold Treatment:** Preconditioning of tissues or embryos at low temperature (usually 4 C) to improve regeneration response or dormancy release.
- **Cotyledon:** The embryonic leaf in seed-bearing plants, often used as an explant source in somatic embryogenesis.
- **Cytokinins:** Plant hormones that promote cell division, shoot formation, and delay senescence (e.g., BAP, Kinetin, TDZ).
- **Differentiation:** The process by which cells or tissues undergo specialization to form organs or embryos.
- **Direct Organogenesis:** Formation of organs (shoots or roots) directly from explants without intervening callus phase.
- **Embryo Rescue:** A technique to culture immature embryos in vitro to bypass post-zygotic incompatibility in crosses.

- **Explant:** A small piece of plant tissue used to initiate tissue culture, such as a leaf disc, shoot tip, or cotyledon.
- **Friable Callus:** A loose, crumbly callus type preferred for establishing suspension cultures.
- **Gibberellins (GA):** A group of plant hormones involved in stem elongation, embryo maturation, and dormancy release.
- **Hyperhydricity (Vitrification):** A physiological disorder in tissue culture resulting in translucent, water-soaked, and fragile plantlets.
- **IBA (Indole-3-butyric acid):** An auxin frequently used in rooting media to promote root development in regenerated shoots.
- **Immersion Frequency:** The interval at which explants are immersed in liquid medium in temporary immersion bioreactors.
- **In Vitro:** Latin for "in glass"; refers to experimental procedures carried out in a controlled, artificial environment such as a culture vessel.
- **Kinetin (KIN):** A synthetic cytokinin often used to stimulate cell division and shoot multiplication.
- **Laminar Airflow Cabinet:** A sterile workbench equipped with HEPA filters, used to perform contamination-free tissue culture work.
- **MS Medium (Murashige and Skoog):** The most widely used basal salt mixture in plant tissue culture.
- **Nodal Segment:** A part of a plant stem containing a node, used as a common explant source for micropropagation.
- **Organogenesis:** The formation of organs from cultured cells or tissues, either directly or via callus.
- **PEG (Polyethylene Glycol):** A fusion-inducing agent used in protoplast fusion protocols.
- **PGRs (Plant Growth Regulators):** Chemical substances (auxins, cytokinins, gibberellins) that regulate plant tissue development in culture.
- **Photoperiod:** The duration of light exposure, often set to 16 hours light/8 hours dark in in vitro plant cultures.
- **Protoplast Fusion:** The joining of two isolated plant protoplasts to produce somatic hybrids with combined genetic traits.
- **Protoplast:** A plant cell with its cell wall removed, used in somatic hybridization and genetic transformation.
- **Regeneration Medium:** Specialized medium containing optimal hormones to regenerate shoots and/or roots from callus or somatic embryos.
- **Shoot Induction:** The initiation of shoot buds from explants, typically triggered by high cytokinin-to-auxin ratios.

- **Shoot Multiplication Ratio:** The number of shoots obtained per explant in a given culture cycle.
- **Somaclonal Variation:** Genetic variability observed among plants regenerated from somatic tissues, often an unintended outcome of tissue culture.
- **Somatic Embryogenesis:** Formation of embryos from somatic or non-reproductive cells, mimicking zygotic embryogenesis.
- **Suspension Culture:** A method of culturing plant cells in a liquid medium under continuous shaking for uniform cell proliferation.
- **Thidiazuron (TDZ):** A powerful cytokinin-like substance that promotes shoot proliferation and somatic embryogenesis in many species.
- **TIB (Temporary Immersion Bioreactor):** A liquid culture system that alternates between immersion and aeration phases to support better growth and reduce hyperhydricity.

Abbreviations and Acronyms

2,4-D	2,4-Dichlorophenoxyacetic acid
BAP	6-Benzylaminopurine
$CaCl_2$	Calcium Chloride
GA_3	Gibberellic Acid
IAA	Indole-3-acetic Acid
IBA	Indole-3-butyric Acid
MS	Murashige and Skoog Medium
NAA	Naphthaleneacetic Acid
PEG	Polyethylene Glycol
PGR	Plant Growth Regulator
pH	Potential of Hydrogen
RITA®	Recipient for Automated Temporary Immersion
SDS	Safety Data Sheet
TIB	Temporary Immersion Bioreactor
TDZ	Thidiazuron
UV	Ultraviolet
NaOCl	Sodium Hypochlorite
EDTA	Ethylenediaminetetraacetic Acid
CH	Casein Hydrolysate

Additional Resources

Here are some excellent resources suitable for both beginners and advanced researchers:

Basic Plant Tissue Culture Techniques (Beginner Level)

- *Plant Tissue Culture: Techniques and Experiments* – Roberta H. Smith
- *Introduction to Plant Tissue Culture* – M.K. Razdan
- *Plant Cell and Tissue Culture – A Tool in Biotechnology* – Karl-Hermann Neumann, Ashwani Kumar, Jánusz I. Szabados

Advanced Tissue Culture Techniques (Intermediate to Expert Level)

- *Plant Tissue Culture, Development, and Biotechnology* – Robert N. Trigiano & Dennis J. Gray.
- *Micropropagation of Orchids* – Joseph Arditti & Robert Ernst
- *Plants from Test Tubes: An Introduction to Micropropagation* – Lydiane Kyte, John Kleyn, Holly Scoggins, and Mark Bridgen

Selected Web Resources:

- Vasil, I. K. (2008). A history of plant biotechnology: from the cell theory of Schleiden and Schwann to biotech crops. *Plant cell reports*, 27, 1423-1440.
- Thorpe, T. A. (2007). History of plant tissue culture. Molecular biotechnology, 37, 169-180.
- Loyola-Vargas, V. M., & Ochoa-Alejo, N. (2018). An introduction to plant tissue culture: advances and perspectives. *Plant cell culture protocols*, 3-13.
- Pasternak, T. P., & Steinmacher, D. (2024). Plant growth regulation in cell and tissue culture in vitro. *Plants*, 13(2), 327.
- Raza, A., et al. (2024). Recent advancements in plant regeneration and tissue culture systems. *Plants*, 13(2), 327.
- Lindsey, K. (Ed.). (2013). Plant Tissue Culture Manual-Supplement 7: Fundamentals and Applications. *Springer Science & Business Media.*
- https://www.ncbi.nlm.nih.gov/pmc/?term=plant+tissue+culture

Disclaimer:

All images included in this manual have been sourced from various publicly available materials on the internet. We acknowledge the original creators and use them here for only illustrative and educational purposes.

INDEX